Legal Notice

\

To inquire about licensing this content, contact the author at

steve@SATPrepGet800.com

BOOKS FROM THE GET 800 COLLECTION
FOR COLLEGE BOUND STUDENTS

The Scholarly Unicorn's SAT Math Advanced Guide
28 SAT Math Lessons to Improve Your Score in One Month
>   Beginner Course
>   Intermediate Course
>   Advanced Course

New SAT Math Problems arranged by Topic and Difficulty Level
320 SAT Math Problems arranged by Topic and Difficulty Level
SAT Verbal Prep Book for Reading and Writing Mastery
320 SAT Math Subject Test Problems
>   Level 1 Test
>   Level 2 Test

320 SAT Chemistry Subject Test Problems
Vocabulary Builder
28 ACT Math Lessons to Improve Your Score in One Month
>   Beginner Course
>   Intermediate Course
>   Advanced Course

320 ACT Math Problems arranged by Topic and Difficulty Level
320 GRE Math Problems arranged by Topic and Difficulty Level
320 AP Calculus AB Problems
320 AP Calculus BC Problems
Physics Mastery for Advanced High School Students
400 SAT Physics Subject Test and AP Physics Problems
SHSAT Verbal Prep Book to Improve Your Score in Two Months
555 Math IQ Questions for Middle School Students
555 Advanced Math Problems for Middle School Students
555 Geometry Problems for High School Students
Algebra Handbook for Gifted Middle School Students
1000 Logic and Reasoning Questions for Gifted and Talented
>   Elementary School Students

CONNECT WITH DR. STEVE WARNER

www.facebook.com/SATPrepGet800

www.youtube.com/TheSATMathPrep

www.twitter.com/SATPrepGet800

www.linkedin.com/in/DrSteveWarner

www.pinterest.com/SATPrepGet800

plus.google.com/+SteveWarnerPhD

The Scholarly Unicorn's
ACT Math Question Bank

Student Workbook with 1000 Problems

Dr. Steve Warner

Table of Contents

1. Purchase a TI-84 or equivalent calculator

It is recommended that you use a TI-84 or comparable calculator for the ACT. Answer explanations (available for free download – see 3 below) will always assume you are using such a calculator.

2. Take a practice ACT from the Official Guide to get your preliminary ACT math score

Use this score to help you determine the problems you should be focusing on. Students scoring below 20 should work on only Level 1, 2, and 3 problems. Students scoring between 20 and 25 should work on Level 1, 2, 3, and 4 problems. Students scoring above 25 should work on all problems.

3. Claim your FREE bonus

See page 219 for details on how to receive solutions to all the problems in this book.

4. 'Like' my Facebook page

This page is updated regularly with standardized test prep advice, tips, tricks, strategies, and practice problems. Visit the following webpage and click the 'like' button.

www.facebook.com/Get800

LESSON 1 – NUMBER THEORY
INTEGERS

Full solutions to all problems in this book are available for free download. See page 219 for details.

LEVEL 1

1. Three consecutive integers are listed in increasing order. If their sum is 54, what is the second integer in the list?

    A. 17
    B. 18
    C. 19
    D. 20
    E. 21

2. If $z + 8$ is an odd integer, then z could be which of the following?

    F. -2
    G. -1
    H. 0
    J. 2
    K. 4

LEVEL 2

3. If z is an odd integer, what is the greatest odd integer less than z ?

    A. $z - 3$
    B. $z - 2$
    C. $z - 1$
    D. $2(z - 1)$
    E. $2(z - 1) - 3$

LEVEL 3

4. If k is an odd integer and m is an even integer, which of the following expressions must be an odd integer?

    F. km
    G. $3km$
    H. $2k + m$
    J. $\frac{km}{2}$
    K. $k^2 + m$

5. Which of the following statements is true about odd and/or even numbers?

    A. The sum of any 2 odd numbers is odd.
    B. The sum of any odd number and any even number is even.
    C. The product of any odd number and any even number is odd.
    D. The product of any 2 odd numbers is odd.
    E. The quotient of any 2 odd numbers is odd.

LEVEL 4

6. If k is a positive integer, which of the following expressions must be an odd integer?

   F. $k + 5$
   G. $\frac{k}{5}$
   H. $5k$
   J. k^5
   K. 5^k

7. The sum of 3 consecutive even integers is m. In terms of m, what is the sum of the 2 greater of these integers?

   A. $m - 3$
   B. $m - 2$
   C. $\frac{2m}{3} - 2$
   D. $\frac{2m}{3}$
   E. $\frac{2m}{3} + 2$

8. The integers 1 through 15 are written on each of fifteen boxes. A boy has placed 44 marbles into these boxes so that the 4th box has the greatest number of marbles. What is the least number of marbles that can be in the 4th box?

   F. 3
   G. 4
   H. 5
   J. 6
   K. 7

LEVEL 5

9. If n is a positive integer such that the units (ones) digit of $n^2 + 4n$ is 7 and the units digit of n is <u>not</u> 7, what is the units digit of $n + 3$?

   A. 1
   B. 2
   C. 3
   D. 4
   E. 5

10. If x and y are integers and $x^2y + xy^2 + x^2y^2$ is odd, which of the following statements must be true?

    I. x is odd
    II. xy is odd
    III. $x + y$ is odd

    F. I only
    G. II only
    H. III only
    J. I and II only
    K. I, II, and III

LESSON 2 – ALGEBRA
SOLVING LINEAR EQUATIONS

LEVEL 1

1. If $z = x - 7$, and $17z - 7z = 30$ what is the value of x?

   A. 2
   B. 4
   C. 6
   D. 8
   E. 10

2. If $8 + x + x = 4 + x + x + x$, what is the value of x?

   F. 1
   G. 2
   H. 3
   J. 4
   K. 5

3. For what value of x is $7x - 6 = 5x + 4$

   A. 10
   B. 5
   C. 1
   D. $\frac{1}{2}$
   E. $\frac{3}{4}$

4. What is the value of x when $\frac{2x}{5} + 12 = 8$

   F. -12
   G. -10
   H. 10
   J. 12
   K. 30

LEVEL 2

5. If $4x - 12 = 8$, then $42 - 3x =$

   A. 5
   B. 12
   C. 20
   D. 27
   E. 35

9

6. If $6(x - 4) = 7(x - 4)$, what is the value of x?

    F. 1
    G. 2
    H. 3
    J. 4
    K. 5

7. The temperature C in degrees Celsius is related to the temperature F in degrees Fahrenheit by the equation $C = \frac{5}{9}(F - 32)$. Which of the following temperatures, in Fahrenheit, is equal to 20 degrees Celsius?

    A. 62
    B. 65
    C. 68
    D. 70
    E. 73

LEVEL 3

8. On Thursday, there are the same number of elephants and giraffes at a zoo. The next day 4 of the elephants are released into the wild leaving twice as many giraffes as elephants at the zoo. How many giraffes are at the zoo?

    F. 4
    G. 6
    H. 8
    J. 12
    K. 16

LEVEL 4

9. There is the same number of cows, pigs and chickens being transported to a farm. When the transport arrives at the farm, 4 cows are taken off the truck and 8 chickens are placed on the truck. If there are now twice as many pigs as cows on the truck, and twice as many chickens as pigs on the truck, how many chickens are on the truck?

    A. 4
    B. 6
    C. 8
    D. 12
    E. 16

LEVEL 5

10. An ornithologist oversees a 400-acre bird sanctuary with only two types of birds: egrets and flamingos. There are currently 150 egrets and 200 flamingos living within the sanctuary. If 75 more egrets are introduced into the sanctuary, how many more flamingos must be introduced so that $\frac{5}{6}$ of the total number of birds in the sanctuary are flamingos?

    F. 925
    G. 650
    H. 425
    J. 350
    K. 150

LESSON 3 – PROBLEM SOLVING AND DATA
RATIOS AND RATES

LEVEL 1

1. The sales tax on a $5.00 hat is $0.40. At this rate, what would be the sales tax on a $9.00 hat?

    A. $0.36
    B. $0.72
    C. $0.80
    D. $0.96
    E. $1.44

LEVEL 2

2. Marco is drawing a time line to represent a 500-year period of time. If he makes the time line 80 inches long and draws it to scale, how many inches will represent each year?

    F. 0.16
    G. 6.25
    H. 6.5
    J. 8.16
    K. 9.05

3. If Robert drove a miles in b hours, which of the following represents his average speed, in miles per hour?

    A. $\frac{a}{b}$
    B. $\frac{b}{a}$
    C. $\frac{1}{ab}$
    D. ab
    E. a^2b

4. The ratio of 17 to 3 is equal to the ratio of 102 to what number?

    F. 578
    G. 36
    H. 18
    J. 2
    K. 0.5

LEVEL 3

5. A kitchen has 1700 square feet of surface that needs to be tiled. If 3 boxes of tiles will cover 710 square feet, what is the least whole number of boxes that must be purchased in order to have enough tiles to cover the entire surface?

    A. 4
    B. 5
    C. 6
    D. 7
    E. 8

11

6. Running at a constant speed, an antelope traveled 150 miles in 6 hours. At this rate, how many miles did the antelope travel in 5 hours?

   F. 80
   G. 100
   H. 112
   J. 125
   K. 180

9. If the ratio of two positive integers is 5 to 4, which of the following statements about these integers CANNOT be true?

   A. Their sum is an odd integer.
   B. Their sum is an even integer.
   C. Their product is divisible by 7.
   D. Their product is an even integer.
   E. Their product is an odd integer.

7. The ratio of the number of elephants to the number of zebras in a zoo is 3 to 5. What percent of the animals in the zoo are zebras?

   A. 12.5%
   B. 37.5%
   C. 60%
   D. 62.5%
   E. 70%

LEVEL 5

10. An elephant traveled 7 miles at an average rate of 4 miles per hour and then traveled the next 7 miles at an average rate of 1 mile per hour. What was the average speed, in miles per hour, of the elephant for the 14 miles?

   F. 1
   G. 1.6
   H. 2.5
   J. 4.375
   K. 5.6

LEVEL 4

8. A mixture is made by combining a green liquid and a purple liquid so that the ratio of the green liquid to the purple liquid is 19 to 6 by weight. How many liters of the purple liquid are needed to make a 370-liter mixture?

   F. 14.8
   G. 29.6
   H. 88.8
   J. 192.4
   K. 281.2

LESSON 4 – GEOMETRY
LINES, ANGLES, AND TRANSFORMATIONS

LEVEL 1

1. In the figure below, four line segments meet at a point to form four angles. What is the value of x ?

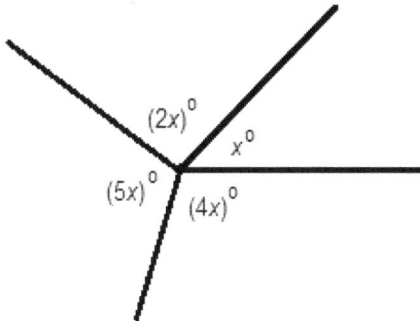

   A. 18
   B. 24
   C. 30
   D. 40
   E. 60

   $(2x)^0$
   x^0
   $(5x)^0$
   $(4x)^0$

2. In the figure below, vertices Q and R of $\triangle QTR$ lie on $\overline{PS}$, the measure of $\angle PQT$ is 125°, and the measure of $\angle QTR$ is 50°. What is the measure of $\angle TRS$?

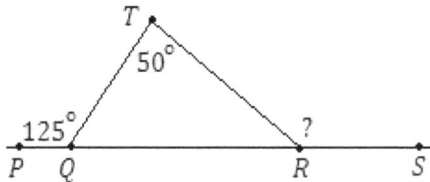

   F. 55
   G. 75
   H. 95
   J. 100
   K. 105

   T
   $50°$
   $125°$
   P Q R S

3. In the standard (x, y) coordinate plane, what is the midpoint of the line segment that has endpoints $(2, -3)$ and $(5, 0)$?

   A. $(-\frac{7}{2}, \frac{3}{2})$
   B. $(-\frac{3}{2}, \frac{7}{2})$
   C. $(-1, -6)$
   D. $(\frac{7}{2}, -\frac{3}{2})$
   E. $(8, 3)$

LEVEL 2

4. In the figure below, what is the value of x ?

   F. 10
   G. 20
   H. 30
   J. 100
   K. 120

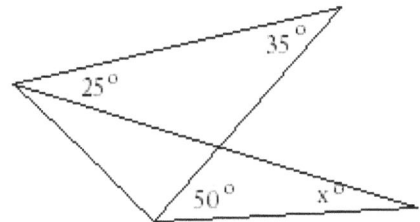

   $35°$
   $25°$
   $50°$
   $x°$

5. C is the midpoint of line segment $\overline{AB}$, and D and E are the midpoints of AC and CB, respectively. If the length of $\overline{AB}$ is 20, what is the length of $\overline{DE}$?

   A. 5
   B. 7.5
   C. 10
   D. 12.5
   E. 15

6. Square $FORM$ with center P is shown below. Point K starts at F and is rotated clockwise about point P a total of $540°$. After the rotation, K is at the same location as which of the following points?

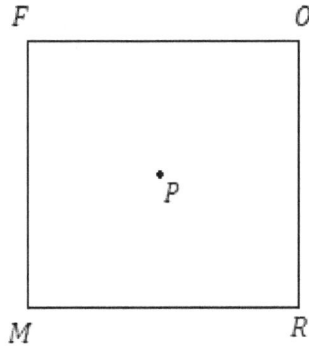

    F. F
    G. O
    H. R
    J. M
    K. P

LEVEL 3

7. In the figure above, $AB = 99$. If $x = \frac{1}{3}y$ and $z = \frac{2}{3}x$, what is the length of line segment $\overline{DB}$?

    A. 11
    B. 22
    C. 33
    D. 34
    E. 35

8. In the figure below, $\overline{AC}$ and $\overline{BD}$ intersect at E, $AE = BE$, $CE = DE$, $m\angle AED = 40°$, and $m\angle BCE = 80°$. What is the measure, in degrees, of $\angle ABC$? (Disregard the degree symbol when gridding your answer.)

    F. 20
    G. 60
    H. 80
    J. 90
    K. 95

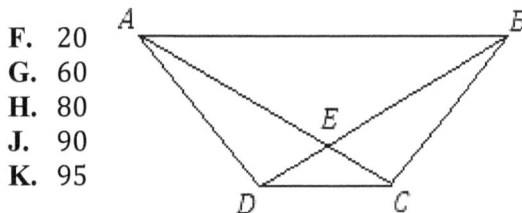

LEVEL 4

9. Consider the transformation of the (x, y) coordinate plane that maps each point (a, b) to the point (ca, cb) for some positive constant c. If this transformation maps $(20, 45)$ to $(4, 9)$, then to where does this transformation map the point $(100, 10)$?

    A. $(20, 2)$
    B. $(95, 5)$
    C. $(90, 3)$
    D. $(10, 4.5)$
    E. $(20, 5)$

LEVEL 5

10. In the figure below, lines k, m, n, and t intersect at a point. If $a + b + c = f + g + h$, which of the following must be true?

    I. $d = e$
    II. $a + b = f + h$
    III. $b + c = g + h$

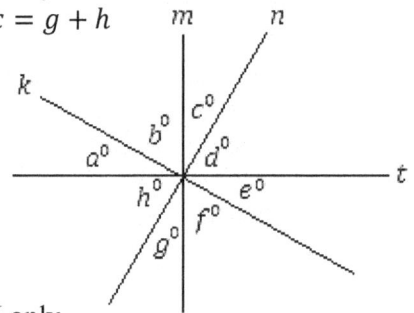

    F. I only
    G. I and II only
    H. I and III only
    J. II and III only
    K. I, II, and III

LESSON 5 – NUMBER THEORY
RATIONAL AND REAL NUMBERS

LEVEL 1

1. The square root of a specific number is approximately 8.72192. The specific number is between what 2 integers?

   A. 2 and 3
   B. 4 and 6
   C. 8 and 16
   D. 22 and 40
   E. 64 and 85

2. Which of the following numbers is less than 0.713 ?

   F. 0.7103
   G. 0.7131
   H. 0.7133
   J. 0.72
   K. 0.721

3. Which of the following numbers has the greatest value?

   A. $0.\overline{7}$
   B. 0.7
   C. 0.77
   D. 0.777
   E. 0.7777

LEVEL 2

4. A piece of cable k feet in length is cut into exactly 4 pieces, each 3 feet 5 inches in length. What is the value of k ?

   F. $12\frac{2}{3}$
   G. 13
   H. $13\frac{1}{3}$
   J. $13\frac{2}{3}$
   K. 14

5. Which of the following is a simplified form of $\sqrt{28} - \sqrt{175}$?

   A. $-3\sqrt{7}$
   B. $-\sqrt{7}$
   C. $\sqrt{7}$
   D. $3\sqrt{7}$
   E. $5\sqrt{7}$

15

LEVEL 3

6. What positive number when divided by its reciprocal has a result of $\frac{9}{16}$?

   F. $\frac{8}{3}$

   G. $\frac{3}{8}$

   H. $\frac{4}{3}$

   J. $\frac{3}{16}$

   K. $\frac{3}{4}$

7. Which of the following number properties is illustrated in the statement below?

$$3(5 + 2) = 3 \cdot 5 + 3 \cdot 2$$

   A. Identity: $1 \cdot x = x$

   B. Inverse: $x \cdot \frac{1}{x} = 1$

   C. Associative: $x(yz) = (xy)z$

   D. Commutative: $xy = yx$

   E. Distributive: $x(y + z) = xy + xz$

LEVEL 4

8. Suppose that k is a positive real number and $\frac{3k^2}{5k^3}$ is a rational number. Which of the following statements about k must be true?

   F. $k = \frac{3}{5}$

   G. $k = 1$

   H. $k = \frac{5}{3}$

   J. k is rational

   K. k is irrational

9. What is one possible value of x for which $\frac{3}{16} < x < \frac{1}{5}$?

   A. $\frac{29}{160}$

   B. $\frac{15}{80}$

   C. $\frac{31}{160}$

   D. $\frac{16}{80}$

   E. $\frac{33}{160}$

LEVEL 5

10. Which of the following expressions equals an irrational number?

   F. $\frac{\sqrt{3}}{\sqrt{27}}$

   G. $\frac{\sqrt{27}}{\sqrt{3}}$

   H. $\sqrt{3} \cdot \sqrt{27}$

   J. $\sqrt{3} + \sqrt{27}$

   K. $\sqrt{3}^2$

LESSON 6 – ALGEBRA
SOLVING LINEAR INEQUALITIES

LEVEL 1

1. For which of the following values of d will the value of $3d - 7$ be greater than 10 ?

    A. 2
    B. 3
    C. 4
    D. 5
    E. 6

2. If $2c + 5 < 27$, which of the following CANNOT be the value of c ?

    F. 7
    G. 8
    H. 9
    J. 10
    K. 11

LEVEL 2

3. Which of the following ordered pairs (x, y) satisfies the inequality $2x - 5y \geq -5$?

    A. $(1, 2)$
    B. $(2, 3)$
    C. $(3, 3)$
    D. $(3, 2)$
    E. $(2, 2)$

4. What is the greatest integer x that satisfies the inequality $3 + \frac{x}{5} < 8$?

    F. 20
    G. 22
    H. 24
    J. 25
    K. 26

LEVEL 3

5. When 11 is decreased by $3x$, the result is more than 5. What is the greatest possible integer value for x ?

    A. 0
    B. 1
    C. 2
    D. 3
    E. 4

$$9 + 3c \geq 8 + 3c$$

6. Which of the following best describes the solutions to the inequality above?

    F. $c \leq 1$
    G. $c \geq 1$
    H. All real numbers
    J. All positive real numbers
    K. No solution

7. Jamie currently has 350 "friends" on a popular social media site. Her goal is to have at least 800 "friends" within the next 15 weeks. What is the minimum number of "friends" per week, on average, she needs to make?

    A. 22
    B. 25
    C. 28
    D. 30
    E. 32

LEVEL 4

$$C = 15h + p + 2000$$

8. The formula above gives the weekly cost C, in dollars, of running a local pizza parlor, where h is the total number of hours the store is open and p is the number of pizzas made. If, during a particular week, the pizza parlor was open for at least 40 hours and it cost no more than $2,750 to run the pizza parlor, what is the maximum number of pizzas that could have been made?

    F. 100
    G. 120
    H. 150
    J. 180
    K. 200

LEVEL 5

9. If $14 \leq x \leq 18$ and $9 \leq y \leq 11$, what is the greatest possible value of $\frac{6}{x-y}$?

    A. $\frac{1}{11}$
    B. $\frac{2}{3}$
    C. $\frac{6}{7}$
    D. $\frac{6}{5}$
    E. 2

10. The daily cost for a publishing company to produce x books is $C(x) = 4x + 800$. The company sells each book for $36. Let $P(x) = R(x) - C(x)$, where $R(x)$ is the total income that the company gets for selling x books. The company takes a loss for the day if $P(x) < 0$. Which of the following inequalities gives all possible integer values of x that guarantee that the company will not take a loss on a given day?

    F. $x > 24$
    G. $x < 24$
    H. $x > 144$
    J. $x < 144$
    K. $x \leq 144$

LESSON 7 – PROBLEM SOLVING AND DATA
PERCENTS

LEVEL 1

1. A box containing 20 ribbons includes 3 red ribbons, 5 blue ribbons, and 12 yellow ribbons. What percent of the ribbons in the box are yellow?

   A. 20%
   B. 40%
   C. 60%
   D. 70%
   E. 80%

2. The price of a house increased from $600,000 to $800,000. The price increased by what percent?

   F. 75%
   G. $66\frac{2}{3}\%$
   H. 50%
   J. $33\frac{1}{3}\%$
   K. 25%

LEVEL 2

3. 10% of 50 is $\frac{1}{6}$ of what number?

   A. $\frac{5}{6}$
   B. 15
   C. 30
   D. 60
   E. 300

4. In September, Daniela was able to type 30 words per minute. In October, she was able to type 42 words per minute. By what percent did Daniela's speed increase from September to October?

   F. 12%
   G. 18%
   H. 30%
   J. 40%
   K. 42%

5. If x is 42% of z and y is 51% of z, what is $x + y$ in terms of z ?

   A. $0.24z$
   B. $0.47z$
   C. $0.78z$
   D. $0.85z$
   E. $0.93z$

LEVEL 3

6. What percent of 75 is 32 ?

   F. $2\frac{11}{32}\%$

   G.   24 %

   H. $37\frac{1}{3}\%$

   J. $42\frac{2}{3}\%$

   K. $234\frac{3}{8}\%$

7. During a sale at a retail store, if a customer buys one t-shirt at full price, the customer is given a 40 percent discount on a second t-shirt of equal or lesser value. If John buys two t-shirts that have full prices of $70 and $90, by what percent is the total cost of the two t-shirts reduced during the sale?

   A. 17.5%
   B. 28%
   C. 40%
   D. 42%
   E. 60%

LEVEL 4

7. If $x > 0$, then 4 percent of 7 percent of $5x$ equals what percent of x ?

   F. 0.14
   G. 1.4
   H. 14
   J. 28
   K. 140

LEVEL 5

9. What is $\frac{1}{7}\%$ of $\frac{14}{3}$?

   A. $\frac{1}{150}$

   B. $\frac{1}{15}$

   C. $\frac{1127}{3500}$

   D. $\frac{49}{150}$

   E. $\frac{2}{3}$

10. If Ted's weight increased by 36 percent and Jessica's weight decreased by 22 percent during a certain year, the ratio of Ted's weight to Jessica's weight at the end of the year was how many times the ratio at the beginning of the year?

   F. $\frac{39}{68}$

   G. $\frac{11}{18}$

   H. $\frac{68}{61}$

   J. $\frac{18}{11}$

   K. $\frac{68}{39}$

LESSON 8 – GEOMETRY
TRIANGLES

LEVEL 1

1. In $\triangle ABC$, the measure of $\angle B$ is 40°, and $\overline{AC} \cong \overline{BC}$, as shown in the figure below. What is the measure of $\angle C$?

   A. 100°
   B. 90°
   C. 80°
   D. 70°
   E. 60°

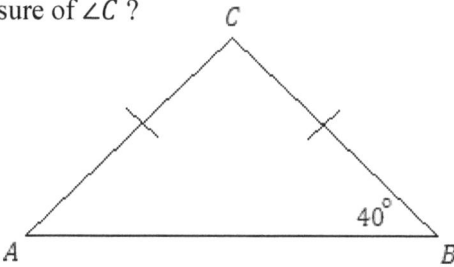

2. In isosceles triangle $\triangle CAT$, $\angle C$ and $\angle A$ are congruent and the measure of $\angle T$ is 52°. What is the measure of $\angle C$?

   F. 148°
   G. 128°
   H. 102°
   J. 86°
   K. 64°

LEVEL 2

3. What is the area of a right triangle whose sides have lengths 6, 8, and 10 ?

   A. 24
   B. 40
   C. 48
   D. 60
   E. 80

4. A right triangle has legs of length 4 cm and 5 cm. The length of the hypotenuse, in centimeters, is between:

   F. 1 and 3
   G. 3 and 4
   H. 4 and 5
   J. 5 and 6
   K. 6 and 8

LEVEL 3

5. In the figure below, what is the area of square $ABCD$?

   A. 9
   B. 81
   C. 106
   D. $\sqrt{56}$
   E. $\sqrt{106}$

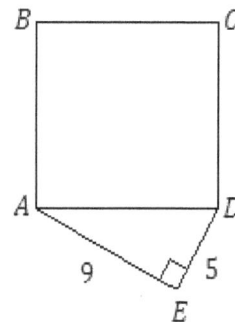

21

6. Which of the following lists of numbers could be the side lengths, in centimeters, of a triangle?

    F. 2, 3, 4
    G. 4, 5, 9
    H. 7, 10, 17
    J. 8, 12, 19
    K. 10, 13, 22

LEVEL 4

7. In the triangle below, $QR = 8$. What is the area of ΔPQR?

    A. $32\sqrt{3}$
    B. 32
    C. $16\sqrt{3}$
    D. 16
    E. $8\sqrt{3}$

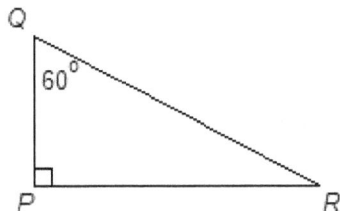

8. What is the area of a square whose diagonal has length $6\sqrt{2}$?

    F. 12
    G. $12\sqrt{2}$
    H. 18
    J. 24
    K. 36

9. A 50-foot ladder is leaning against a vertical wall so that the base of the ladder is 20 feet away from the base of the wall. To the nearest foot, how far up the wall does the ladder reach?

    A. 45
    B. 46
    C. 47
    D. 48
    E. 49

LEVEL 5

10. If x is an integer greater than 8, how many different triangles are there with sides of length 3, 9 and x ?

    F. One
    G. Two
    H. Three
    J. Four
    K. Five

LESSON 9 – NUMBER THEORY
PRIMES, GCD, AND LCM

LEVEL 1

1. Which of the following numbers is NOT a factor of 252 ?

   A. 6
   B. 14
   C. 27
   D. 42
   E. 63

2. What is the least common multiple of 3, 6, 7, 14, and 21 ?

   F. 168
   G. 126
   H. 84
   J. 42
   K. 28

LEVEL 2

3. Given that k is a positive integer and j is 7 times k, what is the least common denominator, in terms of k, for the addition of $\frac{1}{k}$ and $\frac{1}{j}$?

   A. $\frac{k}{7}$
   B. $7k$
   C. $7k^2$
   D. $k + 7$
   E. $k(k + 7)$

4. What is the greatest positive integer that is a divisor of 14, 49, and 63?

   F. 1
   G. 3
   H. 5
   J. 7
   K. 14

LEVEL 3

5. If p is the greatest prime factor of 34 and q is the smallest prime factor of 77, what is the value of pq?

   A. 119
   B. 187
   C. 374
   D. 1309
   E. 2618

6. What is the largest positive integer value of k for which 3^k divides 18^4 ?

    F. 2
    G. 4
    H. 6
    J. 7
    K. 8

7. If an integer n is divisible by 3, 7, 21, and 49, what is the next larger integer divisible by these numbers?

    A. $n + 21$
    B. $n + 49$
    C. $n + 73$
    D. $n + 147$
    E. $n + 294$

8. Let p be a prime number greater than 500,000 and let $z = \frac{1}{\sqrt{p}}$. Which of the following expressions represents a rational number?

    F. $z + 2$
    G. $2z$
    H. $\frac{z}{2}$
    J. $\sqrt{z}$
    K. $3z^2$

LEVEL 4

9. The number 12,121 is the product of the prime numbers 17, 23, and 31. With this knowledge, what is the prime factorization of 145,452 ?

    A. $2 \cdot 3 \cdot 17 \cdot 23 \cdot 31$
    B. $2^2 \cdot 3 \cdot 17 \cdot 23 \cdot 31$
    C. $12 \cdot 17 \cdot 23 \cdot 31$
    D. $3 \cdot 4 \cdot 17 \cdot 23 \cdot 31$
    E. $2^2 \cdot 3^2 \cdot 17 \cdot 23 \cdot 31$

LEVEL 5

10. The integer k is equal to m^2 for some integer m. If k is divisible by 6 and 40, what is the smallest possible positive value of k ?

    F. 720
    G. 900
    H. 1200
    J. 1800
    K. 3600

24

LESSON 10 – ALGEBRA
FUNCTIONS

LEVEL 1

1. Let the function f be defined as
   $f(x) = 3x^2 - 4(5x - 1)$. What is the value of
   $f(-2)$?

   A. -56
   B. -32
   C. $\quad 8$
   D. $\quad 32$
   E. $\quad 56$

LEVEL 2

2. The figure below shows the graph of the
   function f in the standard (xy) coordinate
   plane. What is the value of $f(-2)$?

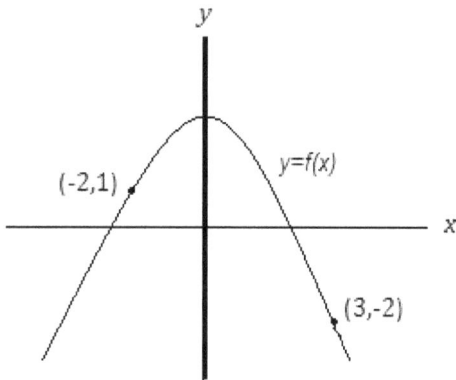

   F. -3
   G. -2
   H. $\quad 1$
   J. $\quad 3$
   K. $\quad 5$

3. Which of the following graphs could not be
   the graph of a function?

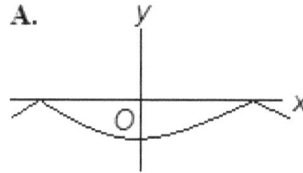

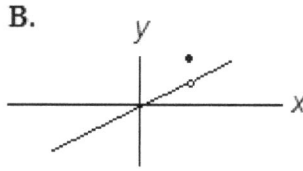

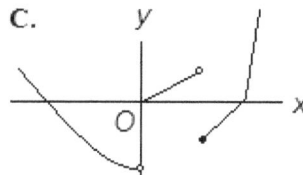

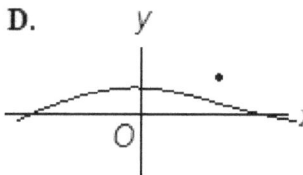

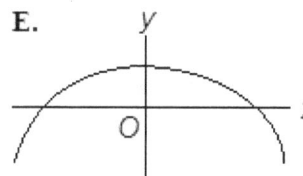

   A.

   B.

   C.

   D.

   E.

4. A function h is defined as
$h(x, y, z) = x^2 - 2yz$. What is $h(-2, 3, 5)$?

   F.   34
   G.   26
   H.   28
   J. -26
   K. -34

LEVEL 3

5. In the xy-plane, the graph of the function f, with equation $f(x) = cx^3 + 5$, passes through the point $(-1, 12)$. What is the value of c ?

   A. -7
   B. -5
   C.   5
   D.   6
   E.   7

6. The figure below shows the graph of the function g. Which of the following is less than $g(1)$?

   F. $g(-3)$
   G. $g(-2)$
   H. $g(-1)$
   J. $g(0)$
   K. $g(3)$

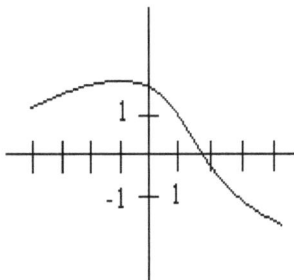

LEVEL 4

x	$p(x)$	$q(x)$	$r(x)$
-2	-3	4	-3
-1	2	1	2
0	5	-1	-6
1	-7	0	-5

7. The functions p, q and r are defined for all values of x, and certain values of those functions are given in the table above. What is the value of $p(-2) + q(0) - r(1)$?

   A.   0
   B.   1
   C.   2
   D.   3
   E.   4

---

8. In the xy-plane below, the graph of the function f is a parabola, and the graph of the function g is a line. The graphs of f and g intersect at $(-2, 1)$ and $(2, -2)$. For which of the following values of x is $f(x) - g(x) < 0$?

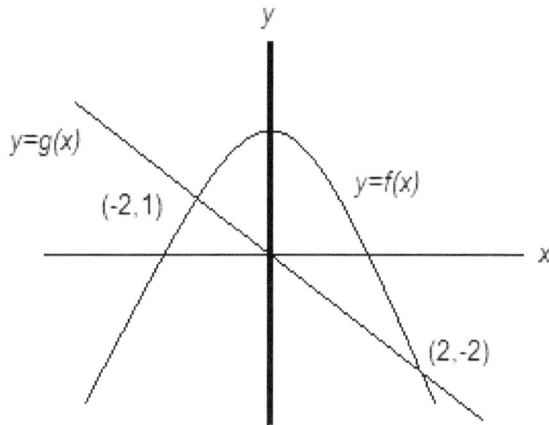

F. -3
G. -1
H. 0
J. 1
K. 2

9. Consider the rational function $r(x) = \frac{x^2-5}{x-3}$. Let $m = r(5)$, let n be the number of horizontal and/or vertical asymptotes there are for the graph of r. What is the value of $m \cdot n$?

A. 10
B. 9
C. 8
D. 7
E. 6

LEVEL 5

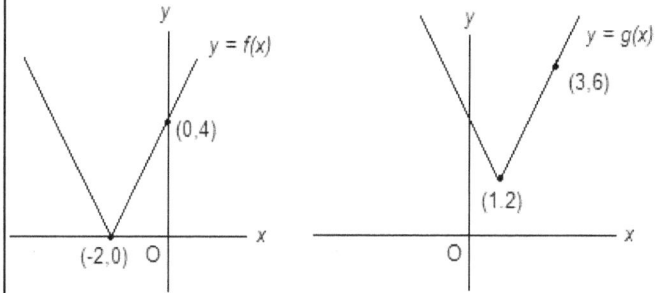

10. The figures above show the graphs of the functions f and g. The function f is defined by $f(x) = 2|x + 2|$ and the function g is defined by $g(x) = f(x + h) + k$, where h and k are constants. What is the value of hk?

F. -6
G. -3
H. -2
J. 3
K. 6

27

LESSON 11 – PROBLEM SOLVING AND DATA STATISTICS

LEVEL 1

1. The average (arithmetic mean) of three numbers is 80. If two of the numbers are 70 and 85, what is the third number?

   A. 75
   B. 80
   C. 85
   D. 90
   E. 95

2. For which of the following lists of 5 numbers is the average (arithmetic mean) less than the median?

   F. 1, 1, 3, 4, 4
   G. 1, 2, 3, 5, 6
   H. 1, 1, 3, 5, 5
   J. 1, 2, 3, 4, 5
   K. 1, 2, 3, 4, 9

3. The average of four numbers is 73. What is the fourth number if the first three of the numbers are 28, 85, and 102 ?

   A. 70
   B. 72
   C. 75
   D. 77
   E. 79

LEVEL 2

4. The average (arithmetic mean) of sixteen numbers is 90. If a seventeenth number, 73, is added to the group, what is the average of the seventeen numbers?

   F. 88
   G. 89
   H. 89.5
   J. 90
   K. 90.5

LEVEL 3

5. The average of a, b, c, and d is 24 and the average of c and d is 14. What is the average of a and b ?

   A. 34
   B. 40
   C. 68
   D. 76
   E. 96

6. The list of numbers 17, A, B, 36, 41, 52 has a mode of 17 and a median of 30. What is the mean of the list, to the nearest whole number?

   F. 28
   G. 30
   H. 31
   J. 32
   K. 34

LEVEL 4

7. If the average (arithmetic mean) of k and $k + 3$ is b and the average of k and $k - 3$ is c, what is the average of b and c?

   A. 1
   B. $\frac{k}{2}$
   C. k
   D. $k + \frac{1}{2}$
   E. $2k$

8. The average (arithmetic mean) of 4 numbers is m. If one of the numbers is n, what is the average of the remaining 3 numbers in terms of m and n?

   F. $\frac{m}{4}$
   G. $4m + n$
   H. $\frac{3m-n}{4}$
   J. $\frac{4m-n}{3}$
   K. $\frac{4n-m}{3}$

LEVEL 5

9. The average (arithmetic mean) salary of employees at an advertising firm with P employees in thousands of dollars is 53, and the average salary of employees at an advertising firm with Q employees in thousands of dollars is 95. When the salaries of both firms are combined, the average salary in thousands of dollars is 83. What is the value of $\frac{P}{Q}$?

   A. $\frac{1}{10}$
   B. $\frac{1}{5}$
   C. $\frac{53}{148}$
   D. $\frac{2}{5}$
   E. $\frac{53}{95}$

$$\frac{1}{x^3}, \frac{1}{x^2}, \frac{1}{x}, x^2, x^3$$

10. If $-1 < x < 0$, what is the median of the five numbers in the list above?

   F. $\frac{1}{x^3}$
   G. $\frac{1}{x^2}$
   H. $\frac{1}{x}$
   J. x^2
   K. x^3

29

LESSON 12 – GEOMETRY
CIRCLES

LEVEL 1

1. The diameter of circle O is 12 inches. What is the circumference of circle O, in inches?

   A. 6π
   B. 12π
   C. 24π
   D. 36π
   E. 144π

2. In the xy-plane, the point $(0,3)$ is the center of a circle that has radius 3. Which of the following is NOT a point on the circle?

   F. $(\ 0,6)$
   G. $(-3,6)$
   H. $(\ 3,3)$
   J. $(-3,3)$
   K. $(\ 0,0)$

LEVEL 2

3. The measure of $\angle ABC$ is 150°. What is the measure of $\angle ABC$ in radians?

   A. $\frac{3}{5\pi}$
   B. $\frac{6}{5\pi}$
   C. $\frac{\pi}{6}$
   D. $\frac{5\pi}{6}$
   E. $\frac{5\pi}{3}$

LEVEL 3

4. What is the total area of the shaded region below to the nearest integer?

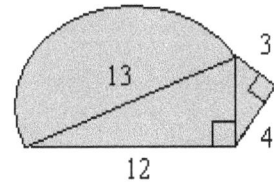

   F. 96
   G. 102
   H. 110
   J. 120
   K. 128

5. What is the area of a circle whose circumference is 18π?

   A. 9π
   B. 24π
   C. 36π
   D. 81π
   E. 243π

LEVEL 4

6. In the figure below, AB is a diameter of the circle with center O and $ABCD$ is a square. What is the area of the shaded region if the radius of the circle is 5?

F. $25(4 - \frac{\pi}{2})$

G. $25(2 - \frac{\pi}{2})$

H. $\pi(4 - \pi)$

J. $\pi(2 - \pi)$

K. $\pi(1 - \pi)$

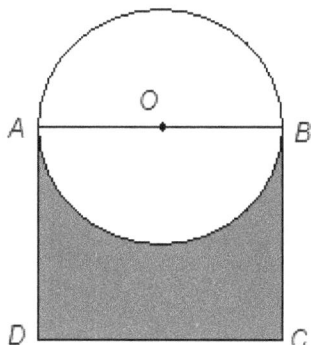

7. Two circles are shown below. The radius of the smaller circle is 3 meters and the radius of the larger circle is 7 meters. What is the area, in square meters, of the shaded region bounded by the circles?

A. 9π

B. 16π

C. 32π

D. 40π

E. 49π

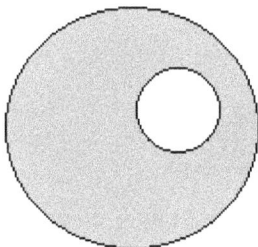

8. In the xy-plane below, O is the center of the circle, and the measure of $\angle POQ$ is $\frac{2\pi}{b}$ radians. What is the value of b ?

F. 2
G. 3
H. 6
J. 8
K. 9

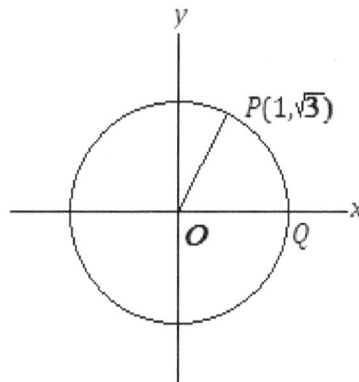

LEVEL 5

9. In the standard (x, y) coordinate plane, what are the coordinates of the center of the circle whose equation is
$x^2 - 8x + y^2 + 10y + 15 = 0$?

A. $(4, 5)$

B. $(4, -5)$

C. $(-4, 0)$

D. $(-4, 5)$

E. $(-5, -4)$

10. How long is the minor axis of the ellipse whose equation is $\frac{(x-3)^2}{25} + \frac{(y+2)^2}{49} = 1$?

F. 5

G. 7

H. 10

J. 14

K. 25

LESSON 13 – NUMBER THEORY
ADDITIONAL PRACTICE 1

LEVEL 1

1. Three consecutive integers are listed in increasing order. If their sum is 138, what is the second integer in the list?

   A. 45
   B. 46
   C. 47
   D. 48
   E. 49

2. Which of the following numbers is between $\frac{1}{9}$ and $\frac{1}{8}$?

   F. 0.10
   G. 0.12
   H. 0.14
   J. 0.16
   K. 0.18

3. What is the greatest common divisor of 14, 49, and 63 ?

   A. 1
   B. 3
   C. 5
   D. 7
   E. 14

LEVEL 2

4. Which of the following groups contains three fractions that are equal?

   F. $\frac{1}{3}, \frac{1}{6}, \frac{1}{9}$
   G. $\frac{2}{3}, \frac{4}{6}, \frac{6}{8}$
   H. $\frac{2}{5}, \frac{4}{25}, \frac{8}{125}$
   J. $\frac{2}{5}, \frac{6}{15}, \frac{10}{25}$
   K. $\frac{2}{5}, \frac{6}{15}, \frac{10}{20}$

5. What is the least common denominator when adding the fractions $\frac{x}{2}, \frac{y}{5}, \frac{z}{25}$, and $\frac{w}{35}$?

   A. 50
   B. 70
   C. 175
   D. 350
   E. 8750

LEVEL 3

6. Which of the following is the decimal equivalent to $\frac{3}{13}$? (Note: A bar indicates a digit pattern that is repeated.)

   F. $0.\overline{230769}$
   G. 0.23076923
   H. $0.230769\overline{23}$
   J. 0.2307692308
   K. $0.2307692\overline{307}$

7. If an integer n is divisible by 21, and 49, what is the next larger integer divisible by these numbers?

    A. $n + 21$
    B. $n + 49$
    C. $n + 70$
    D. $n + 147$
    E. $n + 1029$

10. n is a two-digit number whose units digit is 3 times its tens digit, which of the following statements must be true?

    F. n is less than 15
    G. n is greater than 30
    H. n is a multiple of 3
    J. n is a multiple of 10
    K. n is a multiple of 13

LEVEL 4

8. What is the greatest total number of Sundays there could be in February and March of the same year? (Assume that it is not a leap year so that February has 28 days and March has 31 days).

    F. 8
    G. 9
    H. 10
    J. 11
    K. 12

LEVEL 5

11. If n is a positive integer and $k = n^3 - n$, which of the following statements about k must be true for all values of n?

    I. k is a multiple of 3
    II. k is a multiple of 4
    III. k is a multiple of 6

    A. I only
    B. II only
    C. III only
    D. I and III only
    E. I, II, and III

9. The set P consists of m integers, and the difference between the greatest integer in P and the least integer in P is 455. A new set of m integers, set Q, is formed by multiplying each integer in P by 5 and then adding 3 to the product. What is the difference between the greatest integer in Q and the least integer in Q?

    A. 455
    B. 457
    C. 2275
    D. 2278
    E. 2290

12. The positive number k is the product of four different positive prime numbers. If the sum of these four prime numbers is a prime number greater than 20, what is the least possible value for k?

    F. 354
    G. 390
    H. 462
    J. 490
    K. 770

LESSON 14 – ALGEBRA
ADDITIONAL PRACTICE 1

LEVEL 1

1. If $2j = \frac{x-4}{3}$ and $j = 6$, what is the value of x ?

   A. 10
   B. 20
   C. 30
   D. 40
   E. 50

2. If $3c + 2 < 11$, which of the following CANNOT be the value of c ?

   F. -1
   G. 0
   H. 1
   J. 2
   K. 3

$$g(x) = \frac{2x - 5}{3}$$

3. For the function g above, what is the value of $g(-11)$?

   A. -9
   B. -6
   C. -3
   D. $-\frac{5}{3}$
   E. 9

LEVEL 2

$$\frac{3 + \omega}{2} = 3\frac{1}{2}$$

4. What number, when used in place of ω above, makes the statement true?

   F. 4
   G. 5
   H. 10
   J. 12
   K. 15

5. Given that $a \geq 3$ and $a + b \leq 5$, what is the GREATEST value that b can have?

   A. -8
   B. -2
   C. 0
   D. 2
   E. 8

6. If $f(x) = 2x^3 + 3x - \sqrt{x}$, what is the value of $f(4)$?

   F. 20
   G. 24
   H. 26
   J. 138
   K. 140

34

LEVEL 3

7. For all real numbers x, let the function f be defined as $f(x) = 5x - 10$. Which of the following is equal to $f(3) + f(5)$?

   A. $f(4)$
   B. $f(6)$
   C. $f(7)$
   D. $f(12)$
   E. $f(20)$

LEVEL 4

8. * The monthly membership fee for a fitness center is $49.99. The cost includes the usage of all equipment and classes with the exception of Pilates classes for which there is an additional fee of $1.30 per class. For one month, Cindy's total membership fees were $68.19. How many Pilates classes did Cindy take that month?

   F. 10
   G. 12
   H. 14
   J. 16
   K. 18

9. In the xy-plane, the graph of the function g, with equation $g(x) = px^2 - 25$, passes through the point $(-3, 11)$. What is the value of p ?

   A. -2
   B. 4
   C. 5
   D. 20
   E. 24

10. Let the function g be defined for all real values of x by $g(x) = x(x - 1)$. If a is a positive number and $g(a + 2) = 56$, what is the value of a ?

    F. 1
    G. 2
    H. 3
    J. 6
    K. 8

LEVEL 5

11. * A worker earns $12 per hour for the first 40 hours he works in any given week, and $18 per hour for each hour above 40 that he works each week. If the worker saves 75% of his earnings each week, what is the least number of hours he must work in a week to save at least $441 for the week?

    A. 6
    B. 8
    C. 46
    D. 47
    E. 48

12. Which of the following linear equations gives the vertical asymptote for the graph of $y = \frac{329x + 147}{331x + 149}$?

    F. $x = -\frac{329}{331}$
    G. $x = -\frac{147}{329}$
    H. $x = -\frac{147}{149}$
    J. $x = -\frac{149}{331}$
    K. $x = -\frac{476}{480}$

35

LESSON 15 – PROBLEM SOLVING AND DATA
ADDITIONAL PRACTICE 1

LEVEL 1

1. Alex bought a hat that had an original price of $18.00. The store offered a 25% discount on the original price of the hat, and Alex paid 4% sales tax on the discounted price of the hat. How much did Alex pay for the hat, including tax?

   A. $ 4.68
   B. $12.78
   C. $14.04
   D. $14.22
   E. $14.36

List A: 58, 35, 72, 46, 49
List B: 70, 53, 11, 20, 68

2. The median of the numbers in list B is how much greater than the median of the numbers in list A ?

   F. 2
   G. 4
   H. 7
   J. 7.6
   K. 8

LEVEL 2

3. The ratio of 29 to 5 is equal to the ratio of 203 to what number?

   A.   0.7
   B.   35
   C.   70
   D.   350
   E.   1177.4

4. 30 percent of 50 is 10 percent of what number?

   F. 10
   G. 70
   H. 100
   J. 130
   K. 150

5. For which of the following lists of 6 numbers is the mode NOT equal to the median?

   A. 2, 3, 3, 3, 3, 4
   B. 2, 4, 4, 4, 5, 5
   C. 3, 3, 3, 3, 5, 6
   D. 4, 4, 5, 7, 7, 7
   E. 5, 6, 6, 6, 6, 7

LEVEL 3

6. A bag is filled with black and red marbles so that the ratio of the number of black marbles to the number of red marbles is 2 to 3. Each of the following could be the number of marbles in the bag EXCEPT

   F. 20
   G. 25
   H. 30
   J. 32
   K. 50

36

7. What percent of 60 is 24 ?

   A. 15%
   B. 20%
   C. 40%
   D. 45%
   E. 60%

8. The mean of a list of 10 numbers is 100. A new list of 10 numbers has the same first 8 numbers as the original list, but the ninth number in the new list is 7 more than the ninth number in the old list, and the tenth number in the new list is 2 less than the tenth number in the original list. What is the average of this new list of numbers?

   F. 98
   G. 100
   H. 100.5
   J. 105
   K. 105.5

LEVEL 4

9. If the average (arithmetic mean) of k and $k + 5$ is b and if the average of k and $k - 9$ is c, what is the average of b and c?

   A. $k - 2$
   B. $k - 1$
   C. k
   D. $k + \frac{1}{2}$
   E. $2k$

10. A farmer purchased several animals from a neighboring farmer: 6 animals costing $50 each, 10 animals costing $100 each, and k animals costing $200 each, where k is a positive odd integer. If the median price for all the animals was $100, what is the greatest possible value of k?

    F. 14
    G. 15
    H. 16
    J. 17
    K. 18

LEVEL 5

11. A cheetah ran 12 miles at an average rate of 50 miles per hour and then traveled the next 12 miles at an average rate of 43 mile per hour. What was the average speed, in miles per hour, of the cheetah for the 24 miles? (Round your answer to the nearest tenth)

    A. 47.2
    B. 46.5
    C. 46.4
    D. 46.2
    E. 45.8

12. There are m bricks that need to be stacked. After n of them have been stacked, then in terms of m and n, what percent of the bricks have not yet been stacked?

    F. $\frac{m}{100(m-n)}\%$
    G. $\frac{100(m-n)}{m}\%$
    H. $\frac{100m}{n}\%$
    J. $\frac{100n}{m}\%$
    K. $\frac{m}{100n}\%$

LESSON 16 – GEOMETRY
ADDITIONAL PRACTICE 1

LEVEL 1

1. In the standard (x, y) coordinate plane, what is the midpoint of the line segment that has endpoints $(-2, 5)$ and $(3, -1)$?

   A. $(0, \ 3)$
   B. $(\frac{1}{2}, \ 2)$
   C. $(1, \ 6)$
   D. $(\frac{3}{2}, \ 1)$
   E. $(5, -5)$

2. In the triangle below, $x=$

   F. 62
   G. 64
   H. 66
   J. 68
   K. 70

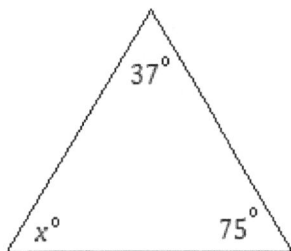

3. In the figure below, A, B, and C lie on the same line. B is the center of the smaller circle, and C is the center of the larger circle. If the diameter of the larger circle is 40, what is the radius of the smaller circle?

   A. 5
   B. 10
   C. 12
   D. 15
   E. 20

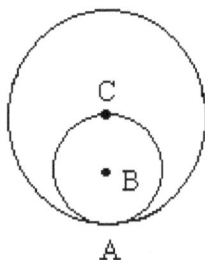

LEVEL 2

4. In the figure below, what is the value of x ?

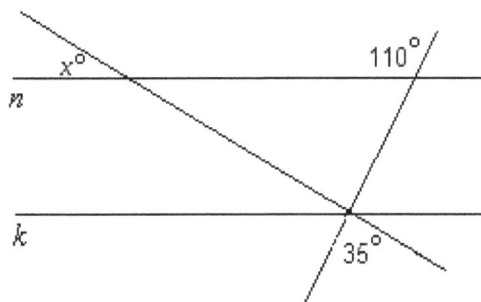

   F. 35
   G. 70
   H. 75
   J. 100
   K. 110

5. In $\triangle ABD$ below, if $y = 42$, what is the value of z ?

   A. 38
   B. 42
   C. 46
   D. 52
   E. 56

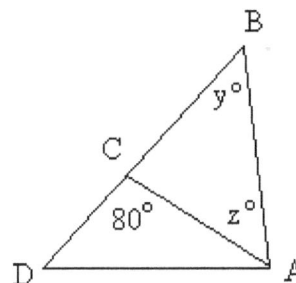

6. What is the radius of a circle whose circumference is π ?

    F. $\frac{1}{2}$
    G. 1
    H. $\frac{\pi}{2}$
    J. π
    K. 2π

LEVEL 3

7. In the figure below, $AB = BC$, $m\angle BCA = 15°$, $m\angle ABD = 10°$, and AC bisects $\angle BCD$. What is $m\angle CED$?

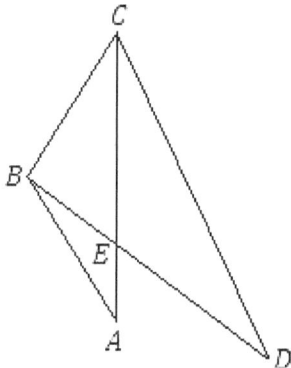

    A. 115°
    B. 128°
    C. 135°
    D. 140°
    E. 155°

8. In the triangle below, what is the length of $\overline{QR}$?

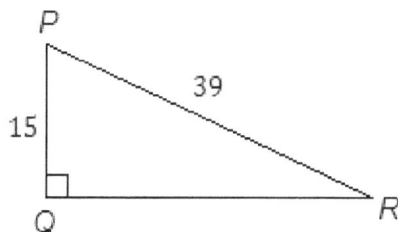

    F. 24
    G. 36
    H. 44
    J. 54
    K. 65

LEVEL 4

9. A triangle ΔPQR, is reflected across the y-axis to have the image $\Delta P'Q'R'$ in the standard (x, y) coordinate plane. So, Q reflects to Q'. If the coordinats of Q are (a, b), what are the coordinates of Q' ?

    A. $(\ a, -b)$
    B. $(-a,\ b)$
    C. $(-a, -b)$
    D. $(\ b,\ a)$
    E. Cannot be determined from the given information

10. In the figure below, each of the points $A, B, C,$ and D is the center of a circle of diameter 6. Each of the four large circles is tangent to two of the other large circles, the small circle, and two sides of the square. What is the length of segment $\overline{BD}$?

    F. $6\sqrt{3}$
    G. $6\sqrt{2}$
    H. $3\sqrt{2}$
    J. 9
    K. 6

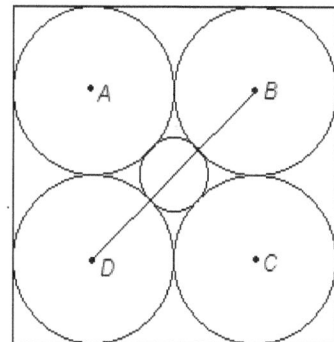

LEVEL 5

11. The lengths of the sides of a triangle are x, 9, and 17, where x is the shortest side. If the triangle is not isosceles, which of the following could be the value of x ?

    A. 7.6
    B. 8
    C. 8.7
    D. 9
    E. 9.1

12. $\overline{AB}$, $\overline{BC}$, and $\overline{AC}$ are diameters of the three circles shown below. If $BC = 4$ and $AB = 5BC$, what is the area of the shaded region?

    F. 48π
    G. 24π
    H. 12π
    J. 6π
    K. 3π

LESSON 17 – NUMBER THEORY
REMAINDERS

LEVEL 1

1. Which of the following numbers disproves the statement "A number that is divisible by 4 and 8 is also divisible by 12" ?

   A. 24
   B. 48
   C. 56
   D. 72
   E. 96

LEVEL 2

2. Starting with a blue tile, colored tiles are placed in a row according to the pattern blue, green, yellow, orange, red, purple. If this pattern is repeated, what is the color of the 73rd tile?

   F. Blue
   G. Green
   H. Yellow
   J. Orange
   K. Red

LEVEL 3

STACKSTAC...

3. In the pattern above, the first letter is S and the letters S, T, A, C, and K repeat continually in that order. What is the 97th letter in the pattern?

   A. S
   B. T
   C. A
   D. D
   E. K

4. Lisa is making a bracelet. She starts with 3 yellow beads, 5 purple beads, and 4 white beads, in that order, and repeats the pattern until there is no more room on the bracelet. If the last bead is purple, which of the following could be the total number of beads on the bracelet?

   F. 81
   G. 85
   H. 87
   J. 88
   K. 93

LEVEL 4

5. Vincent must inspect 9 electronic components that are arranged in a line and labeled numerically from 1 to 9. He must start with component 1 and proceed in order, returning to the beginning and repeating the process after inspecting component 9, stopping when he encounters a defective component. If the first defective component he encounters is component 6, which of the following could be the total number of components that Vincent inspects (counting repetition), including the defective one?

   A. 103
   B. 105
   C. 107
   D. 109
   E. 111

41

6. If k is divided by 9, the remainder is 7. What is the remainder if $4k$ is divided by 6 ?

   F. 5
   G. 4
   H. 3
   J. 2
   K. 1

7. When the positive integer k is divided by 7 the remainder is 4. When the positive integer m is divided by 7 the remainder is 6. What is the remainder when the product km is divided by 7 ?

   A. 1
   B. 2
   C. 3
   D. 4
   E. 5

8. What is the least positive integer greater than 3 that leaves a remainder of 3 when divided by both 6 and 9 ?

   F. 6
   G. 18
   H. 21
   J. 54
   K. 57

LEVEL 5

9. In the repeating decimal
   $0.\overline{123456} = 0.123456123456123 \ldots$
   where the digits 123456 repeat, which digit is in the 2000th place to the right of the decimal?

   A. 1
   B. 2
   C. 3
   D. 4
   E. 5

10. A list consists of 20 consecutive positive integers. Which of the following could be the number of integers in the list that are divisible by 19 ?

    I. None
    II. One
    III. Two

    F. I only
    G. II only
    H. III only
    J. II and III only
    K. I, II, and III

LESSON 18 – ALGEBRA
SOLVING LINEAR SYSTEMS

LEVEL 1

$$y = 3 - x$$
$$2x = 6$$

1. Which of the following ordered pairs (x, y) satisfies the system of equations above?

    A. $(3, -1)$
    B. $(3, \ 0)$
    C. $(3, \ 1)$
    D. $(3, \ 2)$
    E. $(3, \ 3)$

LEVEL 2

2. What is the sum of the solutions of the 2 equations below?

$$2x + 1 = 15$$
$$3y = 11$$

    F. 6
    G. $7\frac{1}{3}$
    H. 8
    J. 10
    K. $10\frac{2}{3}$

3. Which of the following ordered pairs (x, y) satisfies the system of equations below?

$$x + y = 1$$
$$2x + y = 3$$

    A. $(-2, \ 3)$
    B. $(-2, \ 7)$
    C. $(\ 2, \ 3)$
    D. $(\ 2, -1)$
    E. $(\ 2, -2)$

LEVEL 3

4. The system of equations below has one solution (a, b). What is the value of a ?

$$x + 2y = 3$$
$$x + y = 1$$

    F. -2
    G. -1
    H. 0
    J. 3
    K. 7

5. Every Saturday, David spends a total of 6 hours tutoring students. He tutors some students in algebra for 30 minutes each, and he tutors other students in geometry for 45 minutes each. This Saturday, David will tutor twice as many students in geometry as he will tutor in algebra. How many students will he tutor in geometry?

    A. 3
    B. 4
    C. 5
    D. 6
    E. 7

6. There are 19 drivers taking a total of 84 people (including the drivers) on a trip to the museum. Some of the cars can hold 4 people and others can hold 5 people. If all the cars are full, how many cars can hold 5 people?

   F. 4
   G. 5
   H. 6
   J. 7
   K. 8

LEVEL 4

7. Which of the following (x, y) pairs is the solution for the system of equations $\frac{1}{3}x - \frac{1}{6}y = 7$ and $\frac{1}{5}y - \frac{1}{5}x = 8$?

   A. $(-36, -57)$
   B. $(12, 43)$
   C. $(\frac{101}{5}, \frac{307}{5})$
   D. $(82, 122)$
   E. $(122, 82)$

$$3x - 7y = 12$$
$$kx + 21y = -35$$

8. For which of the following values of k will the system of equations above have no solution?

   F. 9
   G. 3
   H. 0
   J. -3
   K. -9

LEVEL 5

9. If $2x = 7 - 3y$ and $5y = 5 - 3x$, what is the value of x?

   A. -10
   B. -5
   C. 10
   D. 20
   E. 25

10. If the system of inequalities $y < 4x + 1$ and $y \geq -\frac{1}{2}x - 2$ is graphed in the xy-plane below, which quadrant contains no solutions to the system?

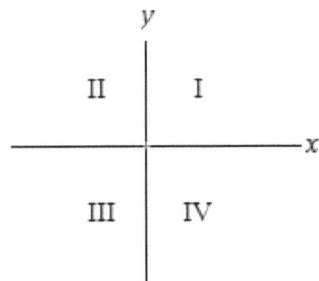

   F. Quadrant I
   G. Quadrant II
   H. Quadrant III
   J. Quadrant IV
   K. There are solutions in all four quadrants.

LESSON 19 – PROBLEM SOLVING AND DATA COUNTING

LEVEL 1

1. A menu lists 2 meals and 3 drinks. How many different meal-drink combinations are possible from this menu?

   A. 5
   B. 6
   C. 8
   D. 9
   E. 12

2. Jamie will be taking history and geometry next semester. Jamie will have 1 of 4 teachers who teach history, and 1 of 3 teachers who teach geometry. From among these 7 teachers, how many possibilities are there for Jamie's 2 teachers for the 2 classes?

   F. 7
   G. 12
   H. 14
   J. 42
   K. 49

LEVEL 2

3. A club consisting of 5 members must choose a president, a vice president, and treasurer. If Joe is chosen to be the treasurer, and no club member can hold more than one position, in how many ways can the president and vice president be chosen?

   A. 12
   B. 16
   C. 25
   D. 60
   E. 120

4. Four pieces of candy will be distributed to 4 children. If each child gets exactly 1 piece of candy and the pieces of candy are all different, in how many ways can the 4 pieces of candy be distributed to the 4 children?

   F. 10
   G. 16
   H. 24
   J. 27
   K. 256

5. Six different books are to be stacked in a pile. In how many different orders can the books be placed on the stack?

   A. 21
   B. 30
   C. 36
   D. 360
   E. 720

45

LEVEL 3

6. While observing several animals in a park, John notices that the squirrel is both the 7th largest and 7th smallest animal. If every animal that John observed was a different size, how many animals did John observe?

   F. 10
   G. 11
   H. 12
   J. 13
   K. 14

7. How many three-digit integers have the hundreds digit equal to 1 or 2, the tens digit equal to 3, 4, or 5 and the units digit (ones digit) equal to 6 or 7 ?

   A. 5
   B. 12
   C. 18
   D. 24
   E. 36

LEVEL 4

8. Jacob will paint his kitchen and bathroom 2 <u>different</u> colors. If 7 different colors are available, how many color combinations are possible for the kitchen and bathroom?

   F. 13
   G. 14
   H. 21
   J. 42
   K. 49

LEVEL 5

9. How many integers between 3000 and 4000 have digits that are all different and that increase from left to right?

   A. 16
   B. 17
   C. 18
   D. 19
   E. 20

10. How many positive integers less than 4000 are multiples of 5 and are equal to 13 times an even integer?

   F. 15
   G. 30
   H. 40
   J. 50
   K. 60

LESSON 20 – GEOMETRY
SOLID GEOMETRY

LEVEL 2

1. What is the volume of a cylinder with base radius 5 cm and height 3 cm?

   A. 15π
   B. 45π
   C. 75π
   D. 225π
   E. 250π

LEVEL 3

2. The volume of a right circular cylinder is 343π cubic centimeters. If the height and base radius of the cylinder are equal, what is the base radius of the cylinder?

   F. 3 centimeters
   G. 5 centimeters
   H. 7 centimeters
   J. 15 centimeters
   K. 25 centimeters

3. How many spherical snowballs with a radius of 4 centimeters can be made with the amount of snow in a spherical snowball of radius 8 centimeters? (the volume V of a sphere with radius r is given by $\frac{4}{3}\pi r^3$.)

   A. 4
   B. 8
   C. 4π
   D. 8π
   E. 12

LEVEL 4

4. In the figure below, segment $\overline{AB}$ joins two vertices of the cube. If the length of $\overline{AB}$ is $3\sqrt{2}$, what is the surface area of the cube?

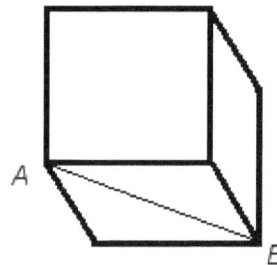

   F. 27
   G. 54
   H. 56
   J. 72
   K. 81

5. A cube of sugar with a side length of 2 inches is placed into a glass of water. The glass is shaped like a right circular cylinder with a diameter of 6 inches and a height of 10 inches. Assuming that the glass is filled to the halfway point with water and the sugar cube is completely submerged in the water, which of the following numerical expressions gives the number of cubic inches of water in the glass?

   A. $\pi(5)^3 - 2^3$
   B. $\pi(3)^2(5) - 2^3$
   C. $\pi(3)(5)^2 - 2^3$
   D. $2\pi(3)(5) - 6(2)^2$
   E. $2\pi(3)(5) + \pi(3)^2(12) - 6(2)^2$

6. The figure below is a right circular cylinder with a height of 10 inches and a base radius of 7 inches. What is the surface area, in square inches, of the cylinder to the nearest integer?

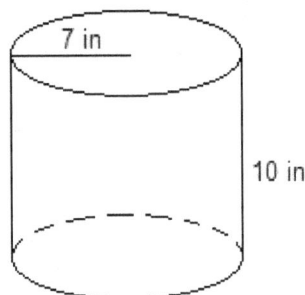

    F.   100
    G. 198
    H. 594
    J.   700
    K. 748

7. Points P and Q are on the surface of a sphere that has a volume of 972π cubic meters. What is the greatest possible length, in meters, of line segment $\overline{PQ}$? (The volume of a sphere with radius r is $V = \frac{4}{3}\pi r^3$.)

    A.   9
    B. 12
    C. 18
    D. 27
    E. 36

LEVEL 5

8. A cube with volume 343 cubic inches is inscribed in a sphere so that each vertex of the cube touches the sphere. What is the length of the radius, in inches, of the sphere?

    F. $\frac{\sqrt{3}}{2}$
    G. $\sqrt{3}$
    H. $\frac{7\sqrt{3}}{2}$
    J. $7\sqrt{3}$
    K. $14\sqrt{3}$

9. How many solid wood cubes, each with a total surface area of 294 square centimeters, can be cut from a solid wood cube with a total surface area of 2,646 square centimeters if no wood is lost in the cutting?

    A.   3
    B.   9
    C. 27
    D. 81
    E. 243

10. The intersection of a plane with a cone could be which of the following?

    I. A circle
    II. A parabola
    III. A trapezoid

    F. I only
    G. II only
    H. I and II only
    J. I and III only
    K. I, II, and III

LESSON 21 – NUMBER THEORY
SCIENTIFIC NOTATION

LEVEL 1

1. The number 720,000,000,000,000 is equivalent to which of the following expressions?

   A. 7.2×10^{15}
   B. 7.2×10^{14}
   C. 7.2×10^{13}
   D. 7.2×10^{-14}
   E. 7.2×10^{-15}

2. The number 0.000 000 000 000 025 is equivalent to which of the following expressions?

   F. 2.5×10^{15}
   G. 2.5×10^{14}
   H. 2.5×10^{-13}
   J. 2.5×10^{-14}
   K. 2.5×10^{-15}

LEVEL 2

3. Danielle's resting heart rate is 65 beats per minute. In scientific notation, how many times would her heart beat in 1 hour, assuming she remains inactive for the full hour?

   A. 6.5×10^1
   B. 65×10^1
   C. 3.9×10^3
   D. 39×10^3
   E. 7.8×10^2

4. In scientific notation, $260,000 + 800,000 = ?$

   F. 1.06×10^{-7}
   G. 1.06×10^{-6}
   H. 1.06×10^5
   J. 1.06×10^6
   K. 1.06×10^7

5. $\frac{9.2 \times 10^{-5}}{2.3 \times 10^{-13}} = ?$

   A. 4.0×10^8
   B. 4.0×10^{-8}
   C. 4.0×10^{-18}
   D. 4.2×10^{18}
   E. 4.2×10^8

LEVEL 3

6. It is estimated that the universe is 13.8 billion years old. When written in scientific notation, which of the following expressions is equal to the number of years used to estimate the age of the universe?

   F. 1.38×10^{10}
   G. 2.76×10^{10}
   H. 1.38×10^9
   J. 13.8×10^9
   K. 27.6×10^9

49

7. The number 0.00007 is 1000 times what number?

   A. 7×10^{-9}
   B. 7×10^{-8}
   C. 7×10^{-7}
   D. 7×10^{8}
   E. 7×10^{9}

9. The number 0.0003 is what number divided by 100 ?

   A. 3×10^{-2}
   B. 3×10^{-1}
   C. 3×10^{0}
   D. 3×10^{1}
   E. 3×10^{2}

LEVEL 4

8. The average distance between Earth and Mars is approximately 225 million kilometers. When written in scientific notation, which of the following expressions is approximately equal to the distance between Earth and Mars, in kilometers?

   F. 2.25×10^{8}
   G. $4.5 \ \times 10^{8}$
   H. 2.25×10^{9}
   J. 22.5×10^{9}
   K. $45 \ \ \times 10^{9}$

10. The radius of the Sun is approximately 4.32×10^{5} miles and the radius of Jupiter is approximately 4.34×10^{4} miles. Which of the following is closest to the difference, in miles, between the diameter of the Sun and the diameter of Jupiter?

   F. 2.0×10^{3}
   G. 2.0×10^{4}
   H. 2.0×10^{5}
   J. 3.9×10^{4}
   K. 3.9×10^{5}

LEVEL 1

1. The expression $x(3 - y)$ is equivalent to:

   A. $3x - y$
   B. $3x + y$
   C. $3x - 3y$
   D. $3x - xy$
   E. $3x - 3xy$

2. $(7c - 8d) + (3d + 4c)$ is equivalent to:

   F. $3c - 5d$
   G. $4c + 4d$
   H. $4c - 4d$
   J. $11c + 11d$
   K. $11c - 5d$

LEVEL 2

3. Which of the following is equivalent to
   $(-3x^2y + 2xy^2) - (-3x^2y - 2xy^2)$?

   A. 0
   B. $-6x^2y$
   C. $4xy^2$
   D. $6x^2y - 4xy^2$
   E. $-6x^2y + 4xy^2$

4. The polynomial $56x^2 + 11x - 15$ is equivalent to the product of $(7x - 3)$ and which of the following binomials?

   F. $49x - 18$
   G. $49x - 12$
   H. $49x + 12$
   J. $8x - 5$
   K. $8x + 5$

5. Which of the following polynomial equations has solutions $-5, 3,$ and 7 ?

   A. $(x - 7)^2(x - 3)(x + 5)^3$
   B. $(x - 5)(x + 5)^2$
   C. $(x - 5)^3(x + 5)$
   D. $(x - 5)(x + 3)(x + 7)$
   E. $(x + 5)^2(x + 3)(x + 7)$

LEVEL 3

6. The expression $(3x - 2)(x + 5)$ is equivalent to:

    F. $3x^2 - 7$
    G. $3x^2 - 10$
    H. $3x - 2x - 7$
    J. $3x^2 + 13x - 10$
    K. $3x^2 - 13x - 10$

7. If $b = 5a^3 - 2a + 7$ and $c = 2a^2 + a + 3$, what is $3c - b$ in terms of a ?

    A. $-5a^3 + 6a^2 + a + 16$
    B. $-5a^3 + 6a^2 + 3a - 4$
    C. $-5a^3 + 6a^2 + 5a + 2$
    D. $a^2 - 5a + 2$
    E. $a^2 + 5a + 2$

LEVEL 4

8. Which of the following is the greatest monomial factor of $54x^2y^3z - 126xy^2z^2$?

    F. $18xyz$
    G. $18xy^2z$
    H. $18x^2y^3z^2$
    J. $378x^2y^3z^3$
    K. $378x^3y^5z^3$

9. For all nonzero values of b, $\dfrac{15b^8 - 20b^5}{5b^2} = ?$

    A. $3b^6 - 4b^3$
    B. $3b^6 - 4b^4$
    C. $3b^6 - 20b^5$
    D. $3b^4 - 4b^3$
    E. $-b^4$

LEVEL 5

10. If $x^2 + y^2 = k^2$, and $xy = 8 - 4k$, what is $(x + y)^2$ in terms of k ?

    F. $k - 4$
    G. $(k - 4)^2$
    H. $k^2 - 4k + 8$
    J. $(k - 2)^2 + 4$
    K. $(k + 4)^2$

LESSON 23 – PROBLEM SOLVING AND DATA PROBABILITY

LEVEL 1

1, 2, 3, 4, 5, 6, 7, 8

1. A number is to be selected at random from the list above. What is the probability that the number selected will be less than 6 ?

   A. $\frac{1}{8}$

   B. $\frac{1}{4}$

   C. $\frac{3}{8}$

   D. $\frac{1}{2}$

   E. $\frac{5}{8}$

2. Stickers in the shape of an isosceles triangle, a hexagon, a parallelogram, and a trapezoid are placed into a bucket. If one of these stickers is taken out at random, what is the probability that the shape chosen will have more than 4 vertices?

   F. $\frac{3}{13}$

   G. $\frac{1}{4}$

   H. $\frac{5}{11}$

   J. $\frac{1}{2}$

   K. $\frac{3}{4}$

3. Patel has decided to donate money to help an animal that requires medical attention. He will choose randomly from 15 animals that are currently being treated by a veterinarian. The number of each type of animal being treated by this veterinarian is listed in the table below. What is the probability that Patel's donation will be used to treat a tiger or an elephant?

Animal	Number needing treatment
Lion	4
Tiger	2
Zebra	5
Elephant	4

   A. $\frac{2}{15}$

   B. $\frac{1}{5}$

   C. $\frac{4}{15}$

   D. $\frac{1}{3}$

   E. $\frac{2}{5}$

4. In a jar, there are exactly 96 coins, of which some are nickels, some are dimes, and the rest are quarters. The probability of randomly selecting a nickel from the jar is $\frac{1}{3}$, and the probability of randomly selecting a dime from the jar is $\frac{1}{6}$. How many coins in the jar are quarters?

   F. 32

   G. 48

   H. 64

   J. 72

   K. 80

LEVEL 2

5. Of the marbles in a jar, 36 are red. Joseph randomly takes one marble out of the jar. If the probability is $\frac{9}{13}$ that the marble he chooses is red, how many marbles are in the jar?

   A. 39
   B. 52
   C. 78
   D. 104
   E. 117

6. Of the 37 marbles in a jar, the most common color is green. What is the probability that a marble randomly selected from the jar is <u>not</u> green?

   F. $\frac{1}{37}$
   G. $\frac{5}{37}$
   H. $\frac{1}{3}$
   J. $\frac{36}{37}$
   K. It cannot be determined from the information given.

LEVEL 3

7. Shown below, a circular board with a spinner has 3 regions (white, black, and grey) whose areas are in the ratio of $1:3:4$, repectively. The spinner is spun and it lands in one of the three regions at random. What is the probability that the region it lands in is the black region?

   A. $\frac{1}{8}$
   B. $\frac{3}{8}$
   C. $\frac{1}{2}$
   D. $\frac{5}{8}$
   E. $\frac{7}{8}$

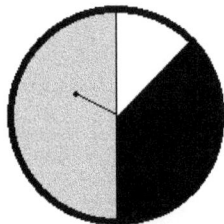

8. A jar contains 11 spheres, 7 cubes, and 4 pyramids. How many additional cubes must be added to the jar so that the probability of randomly selecting a cube is $\frac{6}{11}$?

   F. 8
   G. 11
   H. 14
   J. 17
   K. 23

LEVEL 4

9. If one of the positive factors of 60 is to be chosen at random, what is the probability that the chosen factor will NOT be a multiple of 15?

   A. $\frac{1}{4}$
   B. $\frac{1}{3}$
   C. $\frac{7}{12}$
   D. $\frac{3}{4}$
   E. $\frac{5}{6}$

LEVEL 5

10. The circle graph below shows the distribution of responses to a survey in which a group of men were asked how often they donate to charity. If a man that participated in this survey is selected at random, what is the probability that he donates at least monthly?

    F. 0.33
    G. 0.58
    H. 0.68
    J. 0.73
    K. 0.87

LESSON 24 – GEOMETRY
POLYGONS

LEVEL 1

1. The base and height of a parallelogram, are both 7 inches. What is the area of the parallelogram, in square inches?

   A. 7
   B. 14
   C. 28
   D. 24.5
   E. 49

2. What is the length of a side of a rhombus that has a perimeter of 7 centimeters?

   F. $\sqrt[4]{7}$ centimeters

   G. $\sqrt{7}$ centimeters

   H. $\frac{7}{4}$ centimeters

   J. $\frac{7}{2}$ centimeters

   K. 3 centimeters

LEVEL 2

3. What is the perimeter, in meters, of the figure below?

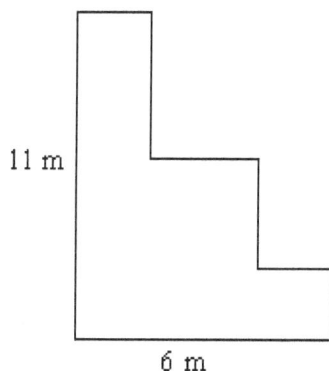

   A. 17
   B. 34
   C. 66
   D. 132
   E. 144

   11 m

   6 m

4. Parallelogram $ABCD$ has a height of 14 meters and an area of 56 square meters. The length of a side of square $PQRS$ is the same as the base of parallelogram $ABCD$. What is the area of square $PQRS$, in square meters?

   F. 4
   G. 8
   H. 12
   J. 16
   K. 20

LEVEL 3

5. If the area of a rectangle is 30 cm then the perimeter of the rectangle, in centimeters, is

   A. 62
   B. 34
   C. 26
   D. 22
   E. Cannot be determined from the given information

6. A trapezoid has a height of 3 inches and an area of 18 square inches. What is the sum of the two bases of the trapezoid, in inches?

    F.   3
    G.   6
    H.   9
    J.   12
    K.   15

8. The consecutive vertices of a certain isosceles trapezoid are P, Q, R, and S where $\overline{PQ} \parallel \overline{SR}$. Which of the following are NOT congruent?

    F. $\angle P$ and $\angle Q$
    G. $\angle R$ and $\angle S$
    H. $\overline{PS}$ and $\overline{QR}$
    J. $\overline{PQ}$ and $\overline{SR}$
    K. $\overline{PR}$ and $\overline{QS}$

LEVEL 4

7. In the figure below, $PQRS$ is a rhombus. Which of the following statements must be true?

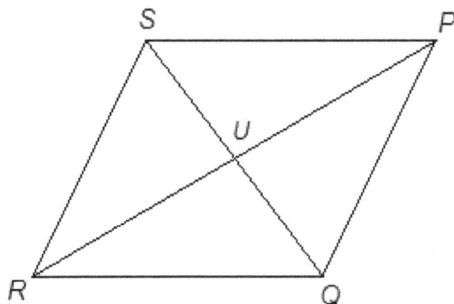

    A. The measure of $\angle RUQ$ is 90°.
    B. The measure of $\angle SUR$ is less than 90°.
    C. The measure of $\angle PQS$ is equal to the measure of $\angle SRP$.
    D. The measure of $\angle RUQ$ is less than the measure of $\angle PQS$.
    E. The measure of $\angle RUQ$ is less than the measure of $\angle SUR$.

LEVEL 5

9. In the figure below, $ABCDEF$ is a regular hexagon and $CD = 6$. What is the perimeter of rectangle $BCEF$ to the nearest tenth?

    A. 1.7
    B. 3.5
    C. 24.0
    D. 28.6
    E. 32.8

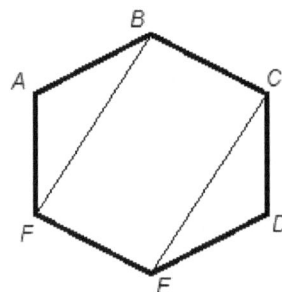

10. How many solid wood cubes, each with a total surface area of 150 square centimeters, can be cut from a solid wood cube with a total surface area of 1,350 square centimeters if no wood is lost in the cutting?

    F.   3
    G.   9
    H. 27
    J. 81
    K. 243

LESSON 25 – NUMBER THEORY
SEQUENCES AND SERIES

LEVEL 1

1. The first term of a sequence is -17. Each term after the first is 8 more than the previous term. What is the first positive number in the sequence?

   A. 1
   B. 2
   C. 3
   D. 5
   E. 7

2. The first term is 3 in the geometric sequence 3, 9, 27,.... What is the FIFTH term of the geometric sequence?

   F. 54
   G. 81
   H. 162
   J. 243
   K. 729

LEVEL 2

20, 59, 30, 89, 45,...

3. In the sequence above, 20 is the first term and each term thereafter is obtained by using the following rules.

   – If the previous term is even, multiply it by 3, then subtract 1.
   – If the previous term is odd, add 1 to it, then divide by 2.

   What is the sixth term of the sequence?

   A. 12
   B. 23
   C. 35
   D. 70
   E. 201

4. The first term of an arithmetic sequence is 7 and the common difference is 3. What is the fifth term of the sequence?

   F. -5
   G. -2
   H. 16
   J. 19
   K. 22

LEVEL 4

5. Each term of a certain sequence is greater than the term before it. The difference between any two consecutive terms in the sequence is always the same number. If the fifth and ninth terms of the sequence are 33 and 97, respectively, what is the twelfth term?

   A. 129
   B. 145
   C. 161
   D. 177
   E. 193

57

6. The first term of the sequence below is 4, and each term after the first is four times the preceding term. Which of the following expressions represents the nth term of the sequence?

$$4, 16, 64, \dots$$

F. $4n$

G. $(n-1)^4$

H. n^4

J. 4^{n-1}

K. 4^n

7. What is the third term of the geometric sequence whose second term is $-\frac{1}{9}$ and whose fifth term is $\frac{1}{243}$?

A. $-\frac{1}{3}$

B. $-\frac{1}{27}$

C. $-\frac{1}{81}$

D. $\frac{1}{27}$

E. $\frac{1}{3}$

$$2, 50, 4, 50, 6, 50, 8, \dots$$

8. In the sequence above, all odd-numbered terms beginning with the first term are the consecutive positive even integers. The even-numbered terms are all 50. What is the difference between the 51st term and the 50th term?

F. 1

G. 2

H. 49

J. 50

K. 51

LEVEL 5

9. If x denotes the sum of the integers from 10 to 70 inclusive, and y denotes the sum of the integers from 80 to 140 inclusive, what is the value of $y - x$?

A. 70

B. 130

C. 700

D. 1400

E. 4270

10. Consecutive terms of an arithmetic sequence have a positive common difference. The sum of the first 4 terms of the sequence is 200. Which of the following values CANNOT be the first term of the arithmetic sequence?

F. 46.5

G. 48

H. 49.5

J. 51

K. None of these can be the first term of the arithmetic sequence.

LESSON 26 – ALGEBRA
ABSOLUTE VALUE

LEVEL 2

1. If $|11 - 3x| > 35$, which of the following is a possible value of x ?

    A. -9
    B. -5
    C. 0
    D. 9
    E. 15

LEVEL 3

2. If $5x - 6 = |-6|$, how many different values are possible for x ?

    F. 0
    G. 1
    H. 2
    J. 3
    K. Infinitely many

3. If the exact weight of an item is X pounds and the estimated weight of the item is Y pounds, then the error, in pounds, is given by $|X - Y|$. Which of the following could be the exact weight, in pounds, of an object with an estimated weight of 6.2 pounds and with an error of less than 0.02 pounds?

    A. 6.215
    B. 6.221
    C. 6.23
    D. 6.3
    E. 6.33

4. Let h be a function such that $h(x) = |5x| + c$ where c is a constant. If $h(4) = -2$, what is the value of $h(-9)$?

    F. -23
    G. -22
    H. 0
    J. 22
    K. 23

5. For all nonzero values of u, v and w, the value of which of the following expressions is *always* negative?

    A. $u - v - w$
    B. $-u - v - w$
    C. $|u| - |v| - |w|$
    D. $-|uv| + |-w|$
    E. $-|u| - |vw|$

LEVEL 4

6. If r and s are nonzero numbers, then which of the following is equivalent to the inequality $|r|\sqrt{7} < |s|\sqrt{2}$?

   F. $r^2 > \frac{2}{7}s^2$

   G. $r^2 < \frac{2}{7}s^2$

   H. $r > \frac{4}{49}s$

   J. $r > \frac{2}{7}s$

   K. $r < \frac{2}{7}s$

7. The number line graph below is the graph of which of the following inequalities?

   A. $|x| \geq 5$
   B. $|x - 1| \geq 5$
   C. $|x - 1| \geq 4$
   D. $|x - 1| \leq 4$
   E. $|x + 1| \geq 4$

8. In a certain game a player can attain a score that is a real number between 0 and 100. The player is said to be in scoring range D if his or her score is between 65 and 83. If John has a score of x, and John is in scoring range D, which of the following represents all possible values of x ?

   F. $|x + 74| < 9$
   G. $|x - 74| < 9$
   H. $|x + 74| = 9$
   J. $|x - 74| > 9$
   K. $|x - 74| = 9$

9. For all $k \neq 0$, $|x| + |y| = -k^2$ has how many (x, y) solutions?

   A. 0
   B. 1
   C. 2
   D. 3
   E. 4

LEVEL 5

10. If $|-3a + 9| = 6$ and $|-2b + 10| = 20$, what is the greatest possible value of ab ?

    F. -25
    G. -5
    H. 15
    J. 75
    K. 77

LESSON 27 – PROBLEM SOLVING AND DATA
CHARTS AND GRAPHS

LEVEL 1

1. The printing cost for customers of Easy Print Shop consists of a fee for printing the document and a price per page for each document. The table below gives the fee and the price per page for customers printing documents of various lengths.

Number of pages	Fee	Price per page
Less than 20	$1.00	$0.10
20 − 100	$2.00	$0.05
More than 100	$3.00	$0.03

Dana wants Easy Print Shop to print a 75 page document for her. What is the total cost for this print job?

A. $2.00
B. $3.25
C. $4.25
D. $5.75
E. $8.50

LEVEL 2

2. The circle graph below describes the time that Juanita spent studying during a Saturday. The number of hours are all correct, but the central angle measures for the sectors are not. What should be the central angle measure for geometry?

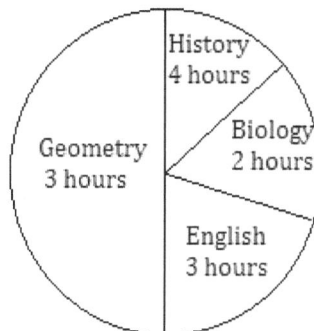

F. 30°
G. 60°
H. 90°
J. 120°
K. 150°

History
4 hours

Biology
2 hours

Geometry
3 hours

English
3 hours

Questions 3 - 4 refer to the following information.

A tracker was implanted inside a bald eagle's wing, and its flight speed was monitored over a period of 2 hours. The data are graphed on the set of axes below with the time elapsed on the   *x*-axis and the flight speed of the eagle on the *y*-axis.

Flight Data from Bald Eagle Tracker

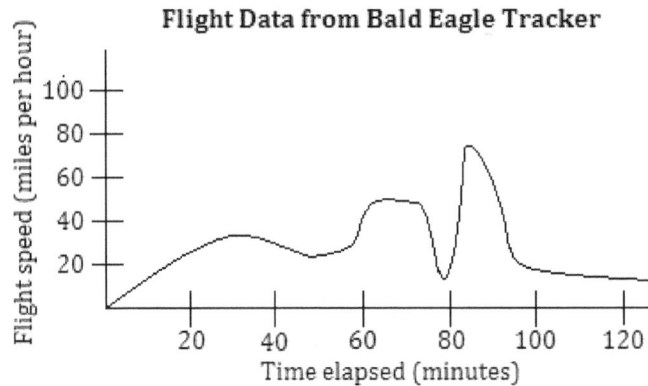

3. On the interval between 75 and 85 minutes, which of the following is closest to the positive difference, in miles per hour, between the greatest and least flight speeds of the bald eagle?

    A. 10
    B. 20
    C. 40
    D. 60
    E. 80

4. On which interval is the eagle's flight speed strictly decreasing then strictly increasing?

    F. Between 0 and 40 minutes
    G. Between 50 and 60 minutes
    H. Between 75 and 85 minutes
    J. Between 90 and 120 minutes
    K. The eagle's flight speed never decreases during the 2 hour time period displayed in the graph.

Questions 5 - 7 refer to the following information.

Gregory has monthly expenses of $4000. These expenses can be summarized in the table below.

Expense	Amount
Rent	$2100
Car Lease	$300
Utilities	$500
Other	$1100

5. Suppose a bar graph will be constructed illustrating the amounts of each expense. The length of the bar for Car Lease should be what fraction of the length of the bar for Rent?

   A. $\frac{3}{40}$
   B. $\frac{1}{7}$
   C. $\frac{1}{6}$
   D. $\frac{1}{5}$
   E. $\frac{11}{40}$

6. In a circle graph illustrating the 4 expenses, what should be the measure of the central angle of the Utilities sector?

   F. 15°
   G. 30°
   H. 45°
   J. 60°
   K. 62.5°

7. The Utilities expense is the sum of the expenses for phone, cable, internet, electricity, and heat. What is the average price per utility?

   A. $250
   B. $200
   C. $150
   D. $120
   E. $100

LEVEL 3

8. According to the line graph below, the mean annual salary of an NBA player in 1981 was what fraction of the mean annual salary of an NBA player in 1984 ?

Mean Annual Salary of NBA Players Each Year from 1980 to 1984

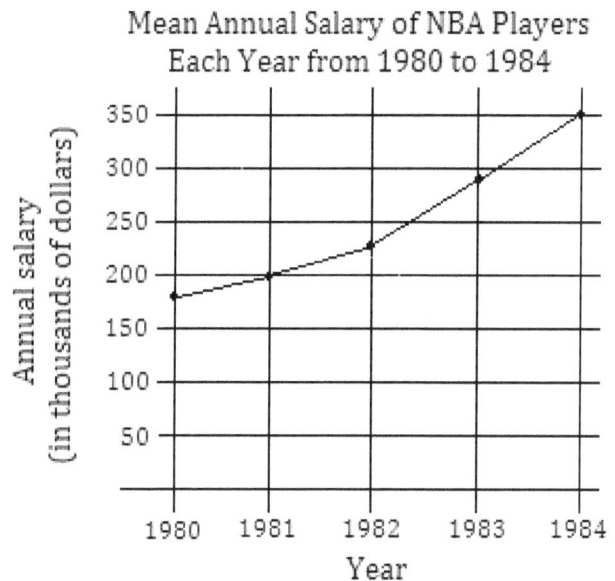

   F. $\frac{3}{7}$
   G. $\frac{4}{7}$
   H. $\frac{9}{14}$
   J. $\frac{5}{7}$
   K. $\frac{11}{14}$

9. The table below lists the results of a survey of a random sample of 500 high school freshman, sophomores and juniors. Each student selected one animal that was his or her favorite.

Favorite Animals

	Dog	Cat	Elephant	Monkey	Lion	Total
Fresh	82	17	20	36	18	173
Soph	51	46	5	50	6	158
Jun	24	30	63	22	30	169
Total	157	93	88	108	54	500

If the sample is representative of a high school with 2,500 freshmen, sophomores and juniors, then based on the table, what is the predicted number of juniors at the high school who would select the elephant as their favorite animal?

A. 63
B. 88
C. 315
D. 580
E. 845

LEVEL 4

10. A quiz was given in a math class, and each student in the class received a whole number score between 70 and 100, inclusive. The frequency chart below shows the cumulative number of students in the math class whose quiz scores fell within certain score ranges. How many students have a quiz score in the interval $81 - 90$?

F. 3
G. 5
H. 8
J. 10
K. 13

Score range	Cumulative number of students
$70 - 80$	10
$70 - 90$	13
$70 - 100$	18

LESSON 28 – GEOMETRY
SLOPE

LEVEL 1

1. What is the slope of the line that passes through the points $(4, -6)$ and $(8, 3)$ in the standard (x, y) coordinate plane?

 A. $-\frac{9}{4}$

 B. $-\frac{4}{9}$

 C. $\frac{4}{9}$

 D. $\frac{9}{10}$

 E. $\frac{9}{4}$

2. What is the equation of line k in the figure below?

 F. $y = -2x + 2$
 G. $y = -2x + 4$
 H. $y = -\frac{1}{2}x + 2$
 J. $y = -\frac{1}{2}x + 4$
 K. $y = 2x + 2$

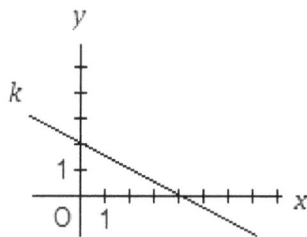

LEVEL 2

3. In the figure below, a line is to be drawn through point A so that it never crosses the x-axis. Through which of the following points must the line pass?

 A. $(\ 7,\ 3)$
 B. $(\ 7, -3)$
 C. $(-7, -3)$
 D. $(\ 3,\ 7)$
 E. $(-3, -7)$

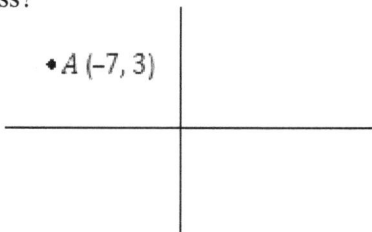

 $\bullet A\,(-7, 3)$

4. In the standard xy-coordinate plane, what is the slope of the line $7x - 3y = 5$?

 F. -7
 G. $\frac{7}{5}$
 H. $\frac{7}{3}$
 J. 5
 K. 7

LEVEL 3

5. Which of the following is an equation of the line in the standard (x, y) coordinate plane that passes through the point $(1, 2)$ and is parallel to the line with equation $y = -2x + 3$?

 A. $2x + y = -4$
 B. $2x + y = -2$
 C. $2x + y = 4$
 D. $-x + 2y = 2$
 E. $-x + 2y = 4$

6. In the figure below, what is the slope of line m?

F. $-\dfrac{1}{2}$

G. $\dfrac{1}{4}$

H. $\dfrac{1}{2}$

J. 2

K. 4

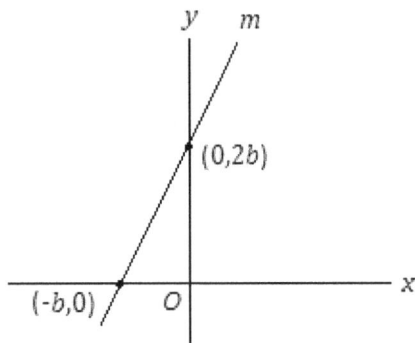

7. In the xy-coordinate plane, line n passes through the points $(0,5)$ and $(-2,0)$. If line m is perpendicular to line n, what is the slope of line m?

A. $-\dfrac{5}{2}$

B. $-\dfrac{2}{5}$

C. 1

D. $\dfrac{2}{5}$

E. $\dfrac{5}{2}$

LEVEL 4

8. The line in the xy-plane below has equation $y = mx + b$, where m and b are constants. What is the value of b?

F. 1

G. $\dfrac{3}{4}$

H. $\dfrac{4}{3}$

J. $\dfrac{8}{3}$

K. 4

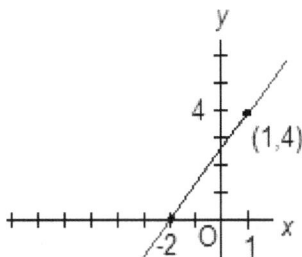

9. The slope of line l in the standard (x, y) coordinate plane is -2. Which of the following is an equation of a line that is perpendicular to line l?

A. $x + 2y = 7$
B. $x - 2y = 7$
C. $2x + y = 7$
D. $2x - y = 7$
E. $-2x + y = 7$

LEVEL 5

10. In the xy-plane, the points $(5, e)$ and $(f, 7)$ are on a line that is perpendicular to the graph of the line $y = -\dfrac{1}{5}x + 12$. Which of the following represents e in terms of f?

F. $5f + 32$
G. $-5f + 32$
H. $5f + 25$
J. $-\dfrac{1}{5}f + 32$
K. $\dfrac{1}{5}f + 32$

LESSON 29 – NUMBER THEORY
ADDITIONAL PRACTICE 2

LEVEL 1

1. How much less than $z + 4$ is $z - 3$?

   A. 1
   B. 3
   C. 4
   D. 7
   E. 12

2. Each of A, B, C, D and E are distinct numbers from the set $\{2, 15, 25, 31, 34\}$ such that A is prime, B is even, C and D are multiples of 5, and $A < E < B$. What is E ?

   F. 2
   G. 15
   H. 25
   J. 31
   K. 34

3. The first term of a sequence is 23. Each term after the first is 7 less than the previous term. What is the first negative number in the sequence?

   A. -1
   B. -2
   C. -3
   D. -5
   E. -7

LEVEL 2

4. What is the least positive integer divisible by the integers 3, 7 and 14 ?

   F. 168
   G. 126
   H. 84
   J. 42
   K. 28

5. In scientific notation,
   $530,000,000 + 900,000,000 = ?$

   A. 1.43×10^{-10}
   B. 1.43×10^{-9}
   C. 1.43×10^{8}
   D. 1.43×10^{9}
   E. 1.43×10^{10}

LEVEL 3

6. What is the largest positive integer value of k for which 7^k divides 147^{15} ?

   F. 3
   G. 7
   H. 15
   J. 28
   K. 30

LEVEL 4

7. For every negative real value of a, all of the following are true EXCEPT:

    A. $|3a| > 0$
    B. $a^7 < 0$
    C. $5a < 0$
    D. $a - |a| = 0$
    E. $4a - 2a^2 < 0$

8. When the positive integer k is divided by 8 the remainder is 5. When the positive integer m is divided by 8 the remainder is 7. What is the remainder when the product km is divided by 4 ?

    F. 0
    G. 1
    H. 2
    J. 3
    K. 4

LEVEL 5

9. If $\frac{jk}{v}$ is an integer which of the following must also be an integer?

    A. jkv
    B. $\frac{3j^2k^2}{v^2}$
    C. $\frac{jv}{k}$
    D. $\frac{kv}{j}$
    E. $\frac{j}{kv}$

10. A business is owned by 1 man and 5 women, each of whom has an equal share. If one of the women sells $\frac{2}{5}$ of her share to the man, and another of the women keeps $\frac{1}{4}$ of her share and sells the rest to the man, what fraction of the business will the man own?

    F. $\frac{9}{40}$
    G. $\frac{37}{120}$
    H. $\frac{2}{3}$
    J. $\frac{43}{120}$
    K. $\frac{3}{8}$

11. If k, m, and n are distinct positive integers such that n is divisible by m, and m is divisible by k, which of the following statements must be true?

    I. n is divisible by k.
    II. $n = mk$.
    III. n has more than 2 positive factors.

    A. I only
    B. III only
    C. I and II only
    D. I and III only
    E. I, II, and III

$$\frac{51}{(2)(3)}, \frac{51}{(3)(4)}, \frac{51}{(4)(5)}, \frac{51}{(5)(6)}$$

12. The first four terms of a sequence are given above. The nth term of the sequence is $\frac{51}{(n+1)(n+2)}$, which is equal to $\frac{51}{n+1} - \frac{51}{n+2}$. What is the sum of the first 100 terms of this sequence?

    F. $\frac{51}{2}$
    G. 25
    H. $\frac{2499}{50}$
    J. $\frac{1224}{50}$
    K. $\frac{1}{50}$

LESSON 30 – ALGEBRA
ADDITIONAL PRACTICE 2

LEVEL 1

1. If $8c + 1 < 25$, which of the following CANNOT be the value of C?

   A. -1
   B. 0
   C. 1
   D. 2
   E. 3

LEVEL 2

2. If $5(x - 7) = 4(x - 8)$, what is the value of x?

   F. 1
   G. 2
   H. 3
   J. 4
   K. 5

3. The figure below shows the graph of the function g in the standard (x, y) coordinate plane. What is the value of $g(b) - g(a)$?

   A. -2
   B. -1
   C. 0
   D. 1
   E. 2

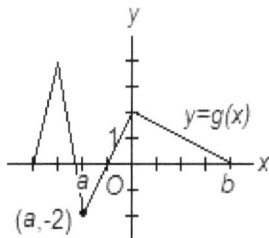

4. The polynomial $56x^2 + 11x - 12$ is equivalent to the product of $(8x - 3)$ and which of the following binomials?

   F. $7x - 9$
   G. $7x + 4$
   H. $7x + 9$
   J. $48x - 9$
   K. $48x + 7$

LEVEL 3

5. Jessica bakes treats for 7 hours every Sunday. It takes her 40 minutes to bake each oatmeal treat and 50 minutes to bake each chocolate treat. This Sunday, Jessica will bake twice as many chocolate treats as oatmeal treats. How many of the chocolate treats will she bake this Sunday?

   A. 3
   B. 4
   C. 5
   D. 6
   E. 8

6. Which of the following is equivalent to $\frac{1}{3}b^3(12a^2 - 5b^4 + 6a^2 + 5b^4)$?

   F. $6a^2b^3$
   G. $18a^2b^4$
   H. $4a^2b^3 + 6a^2$
   J. $6a^2b^3 + 6a^2$
   K. $4a^2b^3 + 6a^2 - 5b^2 + 5b^4$

7. Let h be a function such that $h(x) = |3x| + c$ where c is a constant. If $h(2) = -3$, what is the value of $h(-4)$?

    A. 1
    B. 2
    C. 3
    D. 4
    E. 5

LEVEL 4

8. Let ■ be defined by $x \blacksquare y = \frac{x+y}{x-y}$ for all real numbers x and y, where $x \neq y$. If $1 \blacksquare 2 = 2 \blacksquare k$, what is the value of k ?

    F. 0
    G. 1
    H. 2
    J. 3
    K. 4

9. Which of the following (x, y) pairs is the solution for the system of equations $\frac{1}{3}x - \frac{1}{6}y = 7$ and $\frac{1}{5}y - \frac{1}{5}x = 8$?

    A. $(-36, -57)$
    B. $(12, 43)$
    C. $(\frac{101}{5}, \frac{307}{5})$
    D. $(82, 122)$
    E. $(122, 82)$

LEVEL 5

8. For which of the following functions is it true that $f(-x) = f(x)$ for all values of x ?

    F. $f(x) = x^2 + 5$
    G. $f(x) = x^2 + 5x$
    H. $f(x) = x^3 + 5x$
    J. $f(x) = x^3 + 5$
    K. $f(x) = x + 5$

$$y \leq 2x + 2$$
$$y \geq -3x - 3$$

11. A system of inequalities is shown above, and a graph is shown to the right. Which section or sections of the graph could represent all of the solutions to the system?

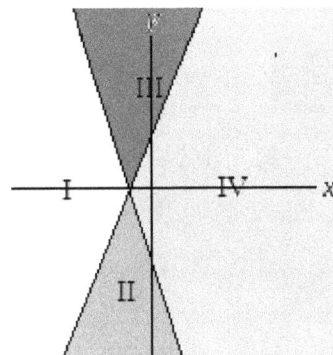

    A. Section I
    B. Section IV
    C. Sections II and III
    D. Sections II and IV
    E. Sections I, II, and IV

12. If $f(x) = x^2 - 5$, which of the following is not true?

    F. $f(-3) = |f(-3)|$
    G. $f(-2) = -|f(2)|$
    H. $f(1) < |f(-1)|$
    J. $f(0) = |f(0)|$
    K. $f(2) < |f(2)|$

LESSON 31 – PROBLEM SOLVING AND DATA
ADDITIONAL PRACTICE 2

LEVEL 1

1. An animal adoption center contains 3 cats, 4 dogs, and 5 rabbits. A family would like to adopt one of each animal. From among these 12 animals, how many different choices does the family have for choosing 1 cat, 1 dog, and 1 rabbit?

   A. 12
   B. 20
   C. 30
   D. 60
   E. 144

2. In a jar there are exactly 72 marbles, each of which is yellow, purple, or blue. The probability of randomly selecting a yellow marble from the jar is $\frac{5}{9}$ and the probability of randomly selecting a purple marble from the jar is $\frac{3}{9}$. How many marbles in the jar are blue?

   F. 8
   G. 24
   H. 32
   J. 40
   K. 64

LEVEL 2

3. If x is 35% of z and y is 60% of z, what is $x + y$ in terms of z ?

   A. 0.21z
   B. 0.45z
   C. 0.75z
   D. 0.81z
   E. 0.95z

4. The average of 3 distinct scores has the same value as the median of the 3 scores. The sum of the 3 scores is 99. What is the sum of the 2 scores that is NOT the median?

   F. 64
   G. 65
   H. 66
   J. 67
   K. 68

5. Six marbles, each of a different color, are to be lined up in a row. In how many different orders can the marbles be arranged?

   A. 720
   B. 540
   C. 36
   D. 30
   E. 21

LEVEL 3

6. What percent of 60 is 12 ?

   F. 10
   G. 12
   H. 15
   J. 18
   K. 20

7. Let m be the median of a set of data containing 13 items. Suppose that six data items are added to the set, three items greater than the original median, and three items less than the original median. Which of the following statements *must* be true about the median of the new data set?

   A. It is less than m.
   B. It is greater than m.
   C. It is equal to m
   D. It is the average of the 3 new lower values.
   E. It is the average of the 3 new higher values.

LEVEL 4

8. If $x \neq 0$ and x is directly proportional to y, which of the following is inversely proportional to $\frac{1}{y^2}$?

   F. x^2
   G. x
   H. $\frac{1}{x}$
   J. $\frac{1}{x^2}$
   K. $-\frac{1}{x^2}$

9. The average (arithmetic mean) age of the people in a certain group was 20 years before one of the members left the group and was replaced by someone who is 10 years older than the person who left. If the average age of the group is now 22 years, how many people are in the group?

   A. 1
   B. 2
   C. 3
   D. 4
   E. 5

10. A jar contains a number of gems of which 75 are blue, 19 are red, and the remainder are white. If the probability of picking a white gem from this jar at random is $\frac{1}{3}$, how many white gems are in the jar?

    F. 6
    G. 25
    H. 32
    J. 47
    K. 63

LEVEL 5

11. Jason ran a race of 1600 meters in two laps of equal distance. His average speeds for the first and second laps were 11 meters per second and 7 meters per second, respectively. To the nearest tenth, what was his average speed for the entire race, in meters per second?

    A. 8.0
    B. 8.6
    C. 9.0
    D. 9.6
    E. 10.0

$$\frac{1}{x^3}, \frac{1}{x^2}, \frac{1}{x}, x, x^2, x^3$$

12. If $-1 < x < 0$, what is the median of the six numbers in the list above?

    F. $\frac{1}{x}$
    G. x^2
    H. $\frac{x^2(x+1)}{2}$
    J. $\frac{x(x^2+1)}{2}$
    K. $\frac{x^2+1}{2x}$

72

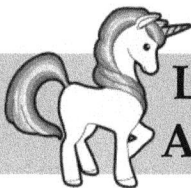

LESSON 32 – GEOMETRY
ADDITIONAL PRACTICE 2

LEVEL 1

1. In the xy-plane, the point $(0, 2)$ is the center of a circle that has radius 2. Which of the following is NOT a point on the circle?

   A. $(0, 4)$
   B. $(-2, 4)$
   C. $(2, 2)$
   D. $(-2, 2)$
   E. $(0, 0)$

LEVEL 2

2. The points P, Q, R, and S are collinear, with Q between P and R and with R between Q and S. Given $PR = 11$ in, $QS = 17$ in, and $QR = 4$ in, what is PS, in inches?

   F. 10
   G. 18
   H. 22
   J. 24
   K. 32

3. In the figure below, one side of a triangle is extended. Which of the following is true?

   A. $y = 75$
   B. $z = 75$
   C. $z - y = 75$
   D. $y + z = 75$
   E. $x = y + z$

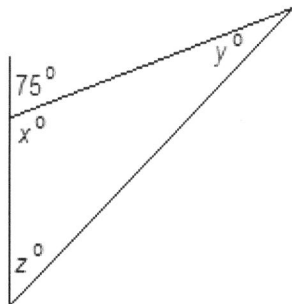

4. If the area of a square is 49 cm^2, then the perimeter of the square, in centimeters, is

   F. 28
   G. 30
   H. 32
   J. 34
   K. 36

LEVEL 3

5. What is the distance between the points $(2, -4)$ and $(-4, 4)$?

   A. 6
   B. 8
   C. 10
   D. 12
   E. 14

6. Line k (not shown) passes through O and intersects $\overline{PQ}$ midway between P and Q. What is the slope of line k?

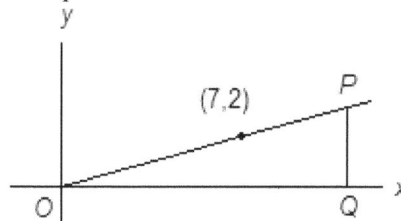

   F. -7
   G. $-\dfrac{1}{7}$
   H. $\dfrac{1}{7}$
   J. 7
   K. Cannot be determined from the given information

73

LEVEL 4

7. The height of a solid cone is 22 centimeters and the radius of the base is 15 centimeters. A cut parallel to the circular base is made completely through the cone so that one of the two resulting solids is a smaller cone. If the radius of the base of the small cone is 5 centimeters, what is the height of the small cone, in centimeters?

   A. $\frac{3}{22}$
   B. 3
   C. 7
   D. 11
   E. $\frac{22}{3}$

8. Point A is a vertex of a 9-sided polygon. The polygon has 9 sides of equal length and 9 angles of equal measure. When all possible diagonals are drawn from point A in the polygon, how many triangles are formed?

   F. One
   G. Three
   H. Five
   J. Seven
   K. Nine

9. If $a > 1$, what is the slope of the line in the xy-plane that passes through the points (a^2, a^4) and (a^3, a^6) ?

   A. $-a^3 + 6a^2$
   B. $-a^3 + a^2$
   C. $-a^3 - a^2$
   D. $a^3 - a^2$
   E. $a^3 + a^2$

LEVEL 5

10. Points $A, B,$ and C lie in a plane. If the distance between A and B is 16 and the distance between B and C is 12, which of the following could NOT be the distance between A and C ?

    I. 3
    II. 27
    III. 28

    F. I only
    G. II only
    H. III only
    J. I and III only
    K. I, II, and III

11. In the figure below, the circle has center O and radius 8. What is the length of arc PRQ ?

    A. 12π
    B. $24\sqrt{2}$
    C. 6π
    D. $12\sqrt{2}$
    E. $3\pi\sqrt{2}$

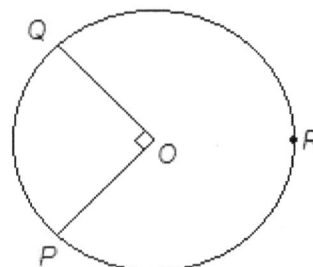

12. A cube is inscribed in a sphere of radius 5 centimeters so that each vertex of the cube touches the sphere. What is the length, in centimeters, of a side of the cube?

    F. 5
    G. $\frac{10\sqrt{3}}{3}$
    H. $\frac{10\sqrt{2}}{2}$
    J. $10\sqrt{2}$
    K. $10\sqrt{3}$

74

LESSON 33 – NUMBER THEORY
COMPLEX NUMBERS

1. For $i = \sqrt{-1}$, the sum $(5 - 9i) + (2 + i)$ is

   A. $3 + 8i$
   B. $3 - 10i$
   C. $-3 - 10i$
   D. $7 - 8i$
   E. $7 - 10i$

LEVEL 2

2. When we subtract $4 - 3i$ from $-2 + 7i$ we get which of the following complex numbers?

   F. $2 - 4i$
   G. $-6 + 4i$
   H. $-6 + 10i$
   J. $6 + 4i$
   K. $6 + 10i$

LEVEL 3

3. Which of the following complex numbers is equivalent to $(5 - 9i)(2 + i)$? (Note: $i = \sqrt{-1}$)

   A. $10 - 9i$
   B. $10 + 9i$
   C. $3 - 8i$
   D. $19 + 13i$
   E. $19 - 13i$

4. For $i = \sqrt{-1}$, $\frac{5}{2+i} \cdot \frac{2-i}{2-i} = ?$

   F. $2 + i$
   G. $2 - i$
   H. $i - 2$
   J. $\frac{2+i}{5}$
   K. $\frac{2-i}{5}$

LEVEL 4

5. If u and v are real numbers, $i = \sqrt{-1}$, and

   $$(v + u) + 6i = 2 - 3ui,$$

   then what is uv ?

   A. -9
   B. -8
   C. 2
   D. 8
   E. 9

6. Which of the following complex numbers equals $(\sqrt{2} + 3i)(5 - i)$?

   F. $5\sqrt{2} - 3i$
   G. $5\sqrt{2} + 3i$
   H. $\left(5\sqrt{2} + 3\right) + \left(15 + \sqrt{2}\right)i$
   J. $\left(5\sqrt{2} + 3\right) + \left(15 - \sqrt{2}\right)i$
   K. $\left(5\sqrt{2} - 3\right) + \left(15 + \sqrt{2}\right)i$

7. If the expression $\frac{5+i}{3-2i}$ is rewritten in the form $a + bi$, where a and b are real numbers, what is the value of $b - a$?

    A. 0
    B. 1
    C. 2
    D. 3
    E. 4

9. For all pairs of nonzero real numbers a and b, the prouct of the complex number $a - bi$ and which of the following complex numbers is a real number?

    A. $a + bi$
    B. $a - bi$
    C. $b + ai$
    D. $b - ai$
    E. $ab + i$

8. After solving a quadratic equation by completing the square, it was found that the equation had solutions, $x = -3 \pm \sqrt{-16b^2}$ where b is a positive real number. Which of the following gives the solutions as complex numbers?

    F. $-3 \pm \quad bi$
    G. $-3 \pm \quad 4bi$
    H. $-3 \pm \quad 8bi$
    J. $-3 \pm \quad 16bi$
    K. $-3 \pm 256bi$

10. If $|z| = 5$, then z CANNOT be which of the following?

    F. -5
    G. $4 - 3i$
    H. $-2 - \sqrt{21}i$
    J. $\sqrt{22} - 3i$
    K. $\sqrt{24} + i$

LESSON 34 – ALGEBRA
MANIPULATING ALGEBRAIC EXPRESSIONS

LEVEL 1

1. If $3y - 18 = 15$, then $y - 6 =$

   A. 30
   B. 20
   C. 15
   D. 10
   E.   5

2. A bank charges a fee of $10 per month to have an account. In addition, there is a charge of $0.05 per check written. Which of the following represents the total charge, in dollars, to have an account for one month in which n checks have been written?

   F. $0.95n$
   G. $1.05n$
   H. $10 + 5n$
   J. $10 + 0.05n$
   K. $10 + 0.05 + n$

LEVEL 2

3. If $\frac{y}{z} = -3$, then $y + 3z =$

   A. -1
   B. 0
   C. 1
   D. y
   E. z

4. Dawn is selling $5d$ CDs at a price of p dollars each. If x is the number of CDs she did <u>not</u> sell, which of the following represents the total dollar amount she received in sales from the CDs?

   F. $px - 5d$
   G. $5d - px$
   H. $5pd - x$
   J. $p(x - 5d)$
   K. $p(5d - x)$

5. Which of the following expressions is equivalent to k less than the product of x and y ?

   A. $x + y - k$
   B. $xy - k$
   C. kxy
   D. $k(x - y)$
   E. $(x - k)y$

LEVEL 3

6. If x is $\frac{3}{5}$ of y and y is $\frac{5}{7}$ of z, what is the value of $\frac{z}{x}$?

  F. $\frac{1}{4}$

  G. $\frac{3}{7}$

  H. $\frac{5}{4}$

  J. $\frac{10}{7}$

  K. $\frac{7}{3}$

7. Bill has cows, pigs and chickens on his farm. The number of chickens he has is three times the number of pigs, and the number of pigs he has is two more than the number of cows. Which of the following could be the total number of these animals?

  A. 14
  B. 15
  C. 16
  D. 17
  E. 18

LEVEL 4

8. For all $x \neq -5$, which of the following expressions is equal to $\frac{x^2+3x-10}{x+5} + 2x - 3$?

  F. $x - 5$
  G. $3x - 5$
  H. $2x^2 - 7x - 6$
  J. $\frac{3x-5}{x+5}$
  K. $\frac{x^2+5x-13}{x+5}$

$$P = \frac{G}{G+N}$$

9. The formula above is used to compute the percentage P of people in any population that play the guitar, where G is the number of people from the population that play the guitar, and N is the number of people from the population that do not play the guitar. Which of the following expresses the number of people that play the guitar in terms of the other variables?

  A. $G = \frac{N}{P-1}$

  B. $G = \frac{N}{1-P}$

  C. $G = \frac{PN}{1-P}$

  D. $G = \frac{PN}{P-1}$

  E. $G = \frac{P}{N-1}$

LEVEL 5

10. For all values of k where the expression is defined, $\dfrac{1}{\frac{1}{k+1}+\frac{1}{k-1}} = ?$

  F. $2k$

  G. $k^2 - 1$

  H. $\frac{1}{k^2-1}$

  J. $\frac{k^2-1}{2k}$

  K. $\frac{2k}{k^2-1}$

LESSON 35 – PROBLEM SOLVING AND DATA
LOGIC AND SETS

LEVEL 2

1. 1000 students were polled to determine which of the following animals they had as pets: cats (C), dogs (D), or birds (B). The Venn diagram above shows the results of the poll. How many students said they have exactly 2 of the 3 types of animals?

   A. 133
   B. 416
   C. 439
   D. 549
   E. 561

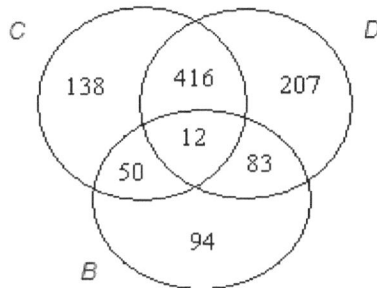

    All of Jim's friends can ski.

2. If the statement above is true, which of the following statements must also be true?

   F. If John cannot ski, then he is not Jim's friend.
   G. If Jeff can ski, then he is not Jim's friend.
   H. If Joseph can ski, then he is Jim's friend.
   J. If James is Jim's friend, then he cannot ski.
   K. If Jordan is not Jim's friend, then he cannot ski.

LEVEL 3

3. Let set A consist of the positive multiples of 15 that are less than 70, and let set B consist of the positive multiples of 9 that are less than 70. How many numbers do sets A and B have in common?

   A. 0
   B. 1
   C. 2
   D. 3
   E. 4

4. In a survey, 62 cat owners were asked about two brands of cat food, Brand X and Brand Y. Of the people surveyed, 26 used Brand X, 11 used Brand Y, and 4 used both brands. How many of the people surveyed didn't use either brand of cat food?

   F. 15
   G. 26
   H. 27
   J. 28
   K. 29

5. Let A and B be two sets of numbers such that every number in B is also in A. Which of the following CANNOT be true?

   A. If 1 is not in A, then 1 is not in B.
   B. 2 is in A, but not in B.
   C. 3 is in B, but not in A.
   D. 4 is in neither A nor B.
   E. 5 is in both A and B.

LEVEL 4

If a beverage is listed in menu A, it is also listed in menu B.

6. If the statement above is true, which of the following statements must also be true?

   F. If a beverage is listed in menu B, it is also in menu A.
   G. If a beverage is not listed in menu A, it is not listed in menu B.
   H. If a beverage is not listed in menu B, it is not listed in menu A.
   J. If a beverage is not listed in menu B, it is in menu A.
   K. If a beverage is listed in menu B, it is not listed in menu A.

Some birds in Bryer Park are ducks.

7. If the statement above is true, which of the following statements must also be true?

   A. Every duck is in Bryer Park
   B. If a bird is not a duck, it is in Bryer Park.
   C. Every bird in Bryer Park is a duck.
   D. All birds in Bryer Park are not ducks.
   E. Not every bird in Bryer Park is not a duck.

8. Set A has a members, set B has b members, and set C consists of all members that are either in set A or set B with the exception of the d members that are common to both ($d >$ 0). Which of the following represents the number of members in set C ?

   F. $a + b + d$
   G. $a + b - d$
   H. $a + b + 2d$
   J. $a + b - 2d$
   K. $2a + 2b - 2d$

9. Which of the following statements is logically equivalent to the following statement?

   If a pig has wings, then it can fly.

   A. If a pig dos not have wings, then it cannot fly.
   B. If a pig cannot fly, then it does not have wings.
   C. A pig has wings if and only if it can fly.
   D. If a pig does not have wings, then it can fly.
   E. If a pig cannot fly, then it does not have wings.

LEVEL 5

10. In Dr. Steve's math class, 10 students have dogs and 15 students have cats. If a total of 19 students have only one of these animals, how many students have both dogs and cats?

    F. 1
    G. 2
    H. 3
    J. 4
    K. 6

LESSON 36 – GEOMETRY
TRIGONOMETRY

LEVEL 2

1. If $0 \leq x \leq 90°$ and $\sin x = \frac{8}{17}$, then $\cos x =$

   A. $\frac{15}{17}$

   B. $\frac{17}{15}$

   C. $\frac{18}{15}$

   D. $\frac{15}{8}$

   E. $\frac{17}{8}$

2. For right triangle ΔPQR shown below, what is $\cos R$?

   F. $\frac{a}{b}$

   G. $\frac{a}{c}$

   H. $\frac{c}{a}$

   J. $\frac{b}{c}$

   K. $\frac{c}{b}$

   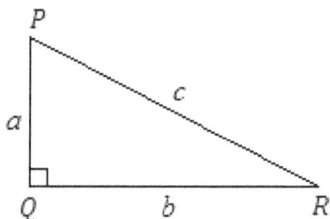

LEVEL 3

3. For right triangle ΔPQR shown below, $\cos P = \frac{4}{11}$. What is the value of $\sin R$?

   A. $\frac{4}{\sqrt{105}}$

   B. $\frac{4}{11}$

   C. $\frac{11}{4}$

   D. $\frac{\sqrt{105}}{11}$

   E. $\frac{11}{\sqrt{105}}$

   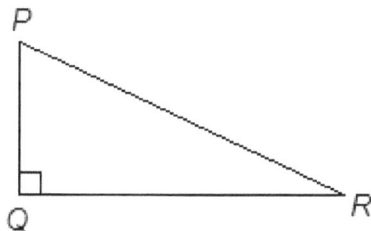

4. The hypotenuse of right triangle ΔABC is 15 inches long, and $\cos A = \frac{1}{5}$. How many inches long is $\overline{BC}$ (note that $\overline{BC}$ is the side opposite $\angle A$) ?

   F. $\frac{1}{6\sqrt{6}}$

   G. $\frac{1}{2\sqrt{6}}$

   H. 3

   J. $2\sqrt{6}$

   K. $6\sqrt{6}$

5. In ΔCAT, $\angle A$ is a right angle. Which of the following is equal to $\tan T$?

   A. $\frac{CA}{CT}$

   B. $\frac{CA}{AT}$

   C. $\frac{CT}{CA}$

   D. $\frac{CT}{AT}$

   E. $\frac{AT}{CA}$

LEVEL 4

6. A 7-foot ladder is leaning against a wall such that the angle relative to the level ground is 70°. Which of the following expressions involving cosine gives the distance, in feet, from the base of the ladder to the wall?

   F. $\dfrac{7}{\cos 70°}$

   G. $\dfrac{\cos 70°}{7}$

   H. $\dfrac{1}{7\cos 70°}$

   J. $7\cos 70°$

   K. $\cos(7 \cdot 70°)$

7. In the triangle below, $QR = 8$. What is the area of $\triangle PQR$?

   A. $32\sqrt{3}$
   B. 32
   C. $16\sqrt{3}$
   D. 16
   E. $8\sqrt{3}$

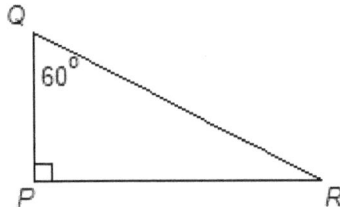

8. Which of the following is equal to $\cos\left(\dfrac{\pi}{5}\right)$?

   F. $-\cos\left(-\dfrac{\pi}{5}\right)$

   G. $-\sin\left(\dfrac{\pi}{5}\right)$

   H. $\sin\left(\dfrac{3\pi}{10}\right)$

   J. $-\sin\left(\dfrac{3\pi}{10}\right)$

   K. $-\cos\left(\dfrac{3\pi}{10}\right)$

LEVEL 5

9. In $\triangle ABC$, the measure of $\angle C$ is 90°, $\tan A = \dfrac{3}{4}$, and $BC = 21$ centimeters. What is the area of $\triangle ABC$, in square centimeters?

   A. 147
   B. 294
   C. 367.5
   D. 490
   E. 735

10. If $\cos x = k$, then for all x in the interval $0 < x < 90°$, $\tan x =$

   F. k

   G. $\dfrac{1}{\sqrt{1-k^2}}$

   H. $\dfrac{k}{\sqrt{1-k^2}}$

   J. $\dfrac{\sqrt{1-k^2}}{k}$

   K. $\sqrt{1-k^2}$

LESSON 37 – NUMBER THEORY
EXPONENTS

LEVEL 1

1. $3a^5 \cdot 5a^7$ is equivalent to:

   A. $8a^{12}$
   B. $8a^{35}$
   C. $15a^2$
   D. $15a^{12}$
   E. $15a^{35}$

2. For what value of x is the equation $7^{5x-2} = 7^{3x}$ true?

   F. 1
   G. 2
   H. 3
   J. 4
   K. 5

LEVEL 2

3. If $6^{x+1} = 7776$, what is the value of x ?

   A. 6
   B. 5
   C. 4
   D. 3
   E. 2

LEVEL 3

4. If $5^x = 7$, then $5^{2x} =$

   F. 11
   G. 13
   H. 25
   J. 49
   K. 625

5. Which of the following expressions is equivalent to $(-x^2y^7)^3$?

   A. $-x^5y^{10}$
   B. $-x^6y^{21}$
   C. $-3x^5y^{10}$
   D. $-3x^6y^{21}$
   E. x^6y^{21}

LEVEL 4

6. Let a and b be nonzero real numbers such that $2^{a+3} = 16b$. Which of the following is an expression for 2^{a+4} in terms of b ?

   F. $\frac{1}{24b^3}$

   G. $\frac{1}{8b}$

   H. b^5

   J. 2^2b^3

   K. 2^5b

$$\sqrt[3]{x^{15}} = x^5$$

7. What are the real number values of x that make the above equation true?

   A. $x < 0$
   B. $x > 0$
   C. $x \leq 0$
   D. $x \geq 0$
   E. All real numbers

8. For all positive real numbers r, which of the following expressions is equivalent to $\dfrac{\left(\frac{x^{32}}{x^8}\right)}{\left(\frac{1}{x^3}\right)}$?

   F. $\sqrt[3]{x^4}$
   G. x^{12}
   H. x^{21}
   J. x^{24}
   K. x^{27}

LEVEL 5

9. Whenever x and y are positive numbers such that $\frac{1}{\sqrt{5}^x} = 25^y$, what is the value of $\frac{y}{x}$?

   A. $-\frac{1}{4}$
   B. -2
   C. -4
   D. 1
   E. 4

10. Seven years ago, Melissa invested $4500 at 5% interest compounded quarterly. Which of the following expressions represents today's value of the investment?

    F. $\$4500e^{0.5t}$
    G. $\$4500(1 + 0.05)^7$
    H. $\$4500\left(1 + \frac{0.05}{4}\right)^{28}$
    J. $\$4500\left(1 + \frac{0.05}{12}\right)^{84}$
    K. $\$4500 + \$4500(4)(7)$

84

LESSON 38 – ALGEBRA
MATRICES

LEVEL 2

1. Which of the following augmented matrices represents the system of linear equations below?

$$4x - y = 7$$
$$3x + 2y = -5$$

A. $\begin{bmatrix} 4 & -1 & | & -7 \\ 3 & 2 & | & 5 \end{bmatrix}$

B. $\begin{bmatrix} 4 & -1 & | & 7 \\ 3 & 2 & | & -5 \end{bmatrix}$

C. $\begin{bmatrix} 4 & 0 & | & 7 \\ 3 & 2 & | & -5 \end{bmatrix}$

D. $\begin{bmatrix} 4 & 1 & | & 7 \\ 3 & 2 & | & -5 \end{bmatrix}$

E. $\begin{bmatrix} 4 & 3 & | & 7 \\ -1 & 2 & | & -5 \end{bmatrix}$

LEVEL 3

2. What is the determinant of the matrix $\begin{bmatrix} 2 & -5 \\ -3 & 6 \end{bmatrix}$?

   F. 27
   G. 16
   H. 0
   J. -3
   K. -27

3. What value of k satisfies the matrix equation below?

$$5\begin{bmatrix} 2 & 0 & 1 \\ 3 & 1 & 2 \end{bmatrix} - 3\begin{bmatrix} 3 & 2 & 0 \\ k & 1 & 2 \end{bmatrix} = \begin{bmatrix} 1 & -6 & 5 \\ 9 & 2 & 4 \end{bmatrix}$$

   A. -2
   B. -1
   C. 0
   D. 1
   E. 2

LEVEL 4

4. Given that $d \begin{bmatrix} 1 & 3 \\ 2 & 4 \end{bmatrix} = \begin{bmatrix} a & b \\ c & 7 \end{bmatrix}$ for some real number d, what is $4ab$?

   F. $\frac{7}{4}$
   G. $\frac{7}{2}$
   H. $\frac{49}{4}$
   J. $\frac{147}{4}$
   K. 49

5. The *determinant* of a matrix $\begin{bmatrix} a & b \\ c & d \end{bmatrix}$ is equal to $ad - bc$. What must be the value of z for the matrix $\begin{bmatrix} z & z \\ z & 6 \end{bmatrix}$ to have a determinant of 9 ?

   A. -6
   B. -3
   C. $-\frac{12}{5}$
   D. $\frac{12}{7}$
   E. 3

6. What is the matrix product $[a \quad b \quad c] \begin{bmatrix} -1 \\ 0 \\ 1 \end{bmatrix}$?

    F. $\quad [-a + c]$

    G. $\quad [-a \quad 0 \quad c]$

    H. $\quad \begin{bmatrix} -a \\ 0 \\ c \end{bmatrix}$

    J. $\quad \begin{bmatrix} -a & -b & -c \\ 0 & 0 & 0 \\ a & b & c \end{bmatrix}$

    K. $\quad \begin{bmatrix} -a & 0 & a \\ -b & 0 & b \\ -c & 0 & c \end{bmatrix}$

LEVEL 5

7. For what positive real value of b, if any, is the determinant of the matrix $\begin{bmatrix} 2 & b \\ b & 7 \end{bmatrix}$ equal to b^2?

    A. 2
    B. 7
    C. $\sqrt{7}$
    D. $\sqrt{14}$
    E. There is no such value of b.

8. Which of the following matrices is equal to the matrix product $\begin{bmatrix} -7 & 3 \\ 2 & -1 \end{bmatrix} \cdot \begin{bmatrix} 2 \\ -3 \end{bmatrix}$?

    F. $\begin{bmatrix} -14 & -9 \\ 4 & 3 \end{bmatrix}$

    G. $\begin{bmatrix} -14 & -9 \\ 3 & 4 \end{bmatrix}$

    H. $\begin{bmatrix} -14 & 4 \\ -6 & 3 \end{bmatrix}$

    J. $\begin{bmatrix} -23 \\ 7 \end{bmatrix}$

    K. $\begin{bmatrix} -5 \\ 1 \end{bmatrix}$

9. Three matrices are given below.

$$A = \begin{bmatrix} 1 & 3 \\ 2 & 4 \end{bmatrix} \quad B = \begin{bmatrix} 4 & 7 \\ 2 & 3 \\ 0 & 1 \end{bmatrix} \quad C = \begin{bmatrix} 1 & 2 & 5 \\ 8 & 4 & 6 \end{bmatrix}$$

Which of the following matrix products is undefined?

    A. AB
    B. AC
    C. BC
    D. BA
    E. BAC

10. Suppose that the determinant of the matrix $\begin{bmatrix} a & b \\ c & d \end{bmatrix}$ equal to k. Which of the following is equal to the determinant of the matrix $\begin{bmatrix} ka & kb \\ hc & hd \end{bmatrix}$?

    F. k
    G. k^2
    H. kh
    J. $k^2 h$
    K. $k^2 + h$

LESSON 39 – PROBLEM SOLVING AND DATA
ADVANCED COUNTING

LEVEL 1

1. Each question on a 3-question quiz offers 4 answers, and exactly 1 answer must be chosen for each question. The quiz has how many possible combinations of answers?

    A. 4
    B. 8
    C. 16
    D. 32
    E. 64

LEVEL 2

2. A chemist is testing 6 different liquids. For each test, the chemist chooses 2 of the liquids and mixes them together. What is the least number of tests that must be done so that every possible combination of liquids is tested?

    F. 36
    G. 30
    H. 15
    J. 12
    K. 8

3. Three light bulbs are placed into three different lamps. How many different arrangements are possible for three light bulbs of different colors – one white, one black, and one yellow?

    A. 6
    B. 9
    C. 18
    D. 27
    E. 30

LEVEL 3

4. Eight different bricks are to be stacked. One brick is chosen for the bottom. In how many different orders can the remaining bricks be stacked?

    F. 27
    G. 35
    H. 1680
    J. 5040
    K. 20,160

5. A committee will be selected from a group of 11 men and 15 women. The committee will consist of 4 men and 6 women. Which of the following expressions gives the number of different committees that can be selected from these 26 people?

    A. $_{26}P_{10}$
    B. $_{26}C_{10}$
    C. $(_{11}P_4)(_{15}P_6)$
    D. $(_{11}C_4)(_{15}C_6)$
    E. $(_{26}C_4)(_{26}C_6)$

LEVEL 4

6. Any 2 points determine a line. If there are 12 points in a plane, no 3 of which lie on the same line, how many lines are determined by pairs of these 12 points?

   F. 14
   G. 24
   H. 66
   J. 108
   K. 132

7. Exactly 5 actors try out for the 5 parts in a play. If each actor can perform any one part and no one will perform more than one part, how many different assignments of actors are possible?

   A. 120
   B. 60
   C. 25
   D. 15
   E. 10

8. Segments $\overline{AC}$, $\overline{AD}$, $\overline{BE}$, and $\overline{EC}$ intersect at the labeled points as shown in the figure below. Define two points as "dependent" if they lie on the same segment in the figure. Of the labeled points in the figure, how many pairs of dependent points are there?

   F. None
   G. Three
   H. Six
   J. Nine
   K. Twelve

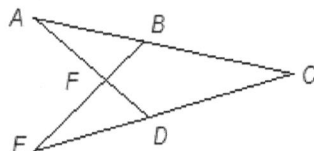

LEVEL 5

9. If seven cards, each of a different color are placed in a row so that the green one is placed at an end, how many different arrangements are possible?

   A. 7
   B. 42
   C. 720
   D. 1440
   E. 9030

10. Regular pentagons have 5 diagonals as shown below. How many diagonals does a regular octagon have? (An octagon is a polygon with 8 sides)

    F. 40
    G. 30
    H. 20
    J. 16
    K. 8

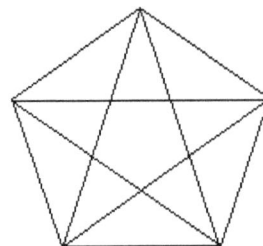

LESSON 40 – GEOMETRY
PARALLEL LINES AND SIMILARITY

LEVEL 2

1. In the figure below, ℓ and m are parallel. Which of the following equations must be true?

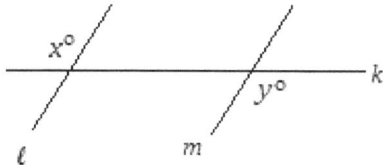

   A. $x + y = 90$
   B. $x + y = 180$
   C. $x + y = 360$
   D. $x = y$
   E. $x = 2y$

2. If the two triangles in the figure below are similar, what is the value of h ?

   F. 54
   G. 45
   H. 42
   J. 36
   K. 30

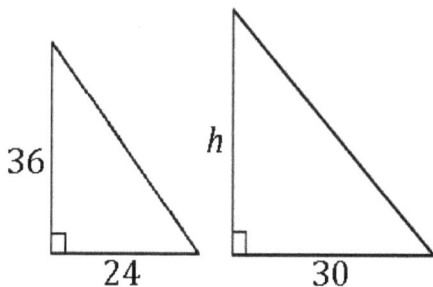

3. In the figure below, $AE \parallel CD$ and segment AD intersects segment CE at B. What is the length of segment CE ?

   A. 3
   B. 4
   C. 7
   D. 11
   E. 12

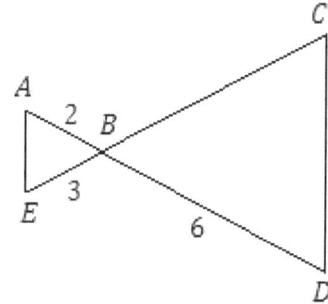

4. In the figure below, lines j and k are parallel and lines ℓ and m are parallel. If the measure of $\angle 1$ is 55°, what is the measure of $\angle 2$?

   F. 35°
   G. 55°
   H. 105°
   J. 125°
   K. 155°

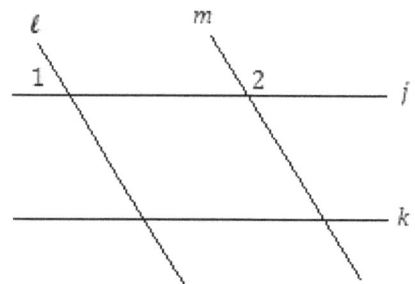

LEVEL 3

5. In the figure below, what is the value of $\frac{ED}{AD}$?

    A. $\frac{1}{7}$

    B. $\frac{1}{4}$

    C. $\frac{2}{5}$

    D. $\frac{1}{2}$

    E. $\frac{6}{7}$

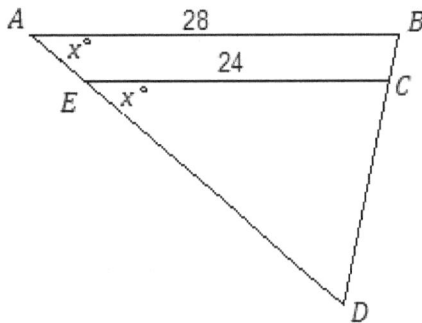

6. In the figure below, line ℓ is parallel to line k. Transversals m and n intersect at point P on ℓ and intersect k at points R and Q, respectively. Point Y is on k, the measure of $\angle PRY$ is 140°, and the measure of $\angle QPR$ is 100°. How many of the angles formed by rays ℓ, k, m, and n have measure 40° ?

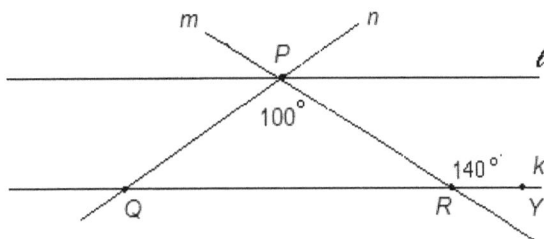

    F. 4
    G. 6
    H. 8
    J. 10
    K. 12

7. A 36 feet tall tree is casting a shadow 24 feet long. At the same time, a nearby tree is casting a shadow 30 feet long. If the lengths of the shadows are proportional to the heights of the trees, what is the height, in feet, of the taller tree?

    A. 54
    B. 45
    C. 42
    D. 36
    E. 30

LEVEL 4

8. Given the two similar right triangles below with dimensions given in centimeters, what is the area, in square centimeters of the larger triangle?

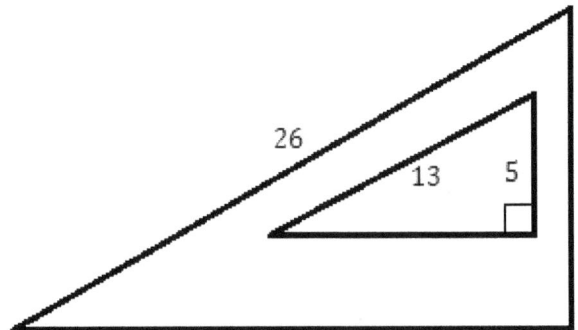

    F. 60
    G. 80
    H. 100
    J. 120
    K. 240

90

9. The ratio of the side lengths for a triangle is $3:7:14$. In a second triangle similar to the first, the longest side is 11 feet. To the nearest tenth of a foot, what is the length of the shortest side of the second triangle?

    A. 2.2
    B. 2.4
    C. 3.6
    D. 5.5
    E. Cannot be determined from the given information

LEVEL 5

10. In the triangle below, $DC = 3$ and $BC = 6$. What is the value of AC ?

    F. 3
    G. 6
    H. 9
    J. 12
    K. 15

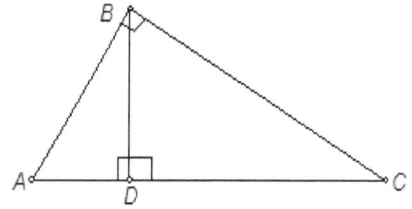

LESSON 41 – NUMBER THEORY
LOGARITHMS

LEVEL 2

1. What is the value of $\log_3 3$?

   A. 0
   B. 1
   C. 3
   D. 6
   E. 9

LEVEL 3

2. In the equation $\log_3 45 - \log_3 5 = \log_3 x$, the value of x is ?

   F. 6
   G. 7
   H. 8
   J. 9
   K. 10

LEVEL 4

3. What is the value of $\log_5 125$?

   A. 3
   B. 4
   C. 6
   D. 10
   E. 16

4. Whenever n is an integer greater than 1, $\log_n n^3 n^5 = ?$

   F. 2
   G. 8
   H. 12
   J. 15
   K. 30

5. The value of $\log_3\left(3^{\frac{21}{4}}\right)$ is between which of the following pairs of consecutive integers?

   A. 0 and 1
   B. 2 and 3
   C. 4 and 5
   D. 5 and 6
   E. 6 and 7

6. For all $x > 0$, which of the following expressions is equivalent to $\log\left((6x)^{\frac{1}{3}}\right)$?

   F. $\log 2x$
   G. $\log 2 + \log \frac{x}{3}$
   H. $\log 6 + \frac{1}{3}\log x$
   J. $\frac{1}{3}\log 6 + \frac{1}{3}\log x$
   K. $\frac{1}{3}(\log 6)(\log x)$

7. What is the value of $\log_3 81$?

   A. 3
   B. 4
   C. 8
   D. 9
   E. 27

LEVEL 5

8. In the equation $\log_4 2 + \log_4 8 = \log_6 x^2$, what is the positive real value of x ?

   F. 6
   G. 7
   H. 8
   J. 9
   K. 10

9. For what real value of x, if any, is $\log_{(x+5)}(x^2 + 5) = 2$?

   A. -3
   B. -2
   C. -1
   D. 0
   E. There is no such value of x.

10. The magnitude of an earthquake, M, can be modeled by the equation $M = \log \frac{I}{S}$, where I is the intensity of the earthquake and S is a constant. What is the magnitude of an earthquake whose intensity is 10,000 times the value of S ?

   F. 1
   G. 4
   H. 10
   J. 30
   K. 100

LESSON 42 – ALGEBRA
QUADRATIC EQUATIONS

LEVEL 2

1. If $(x - 3)^2 = 36$, and $x < 0$, what is the value of x?

   A. -33
   B. -9
   C. -3
   D. -2
   E. -1

LEVEL 3

2. For what two values of x is the equation $x^2 + 3x - 10 = 0$ true?

   F. -2 and -5
   G. 2 and -5
   H. 2 and 5
   J. -2 and 5
   K. -2 and 2

3. In the quadratic equation $x^2 - 2x = 15$, find the positive solution for x.

   A. 1
   B. 2
   C. 3
   D. 4
   E. 5

4. Which of the following describes the nature of the roots of the equation $x^2 + 8x + 7 = 0$?

   F. real, rational, unequal
   G. real, irrational, unequal
   H. real, rational, equal
   J. real, irrational, equal
   K. complex

5. What is the sum and product of the two solutions of the equation $x^2 - x + 15 = 0$?

   A. sum $= -1$, product $= 15$
   B. sum $= 1$, product $= 15$
   C. sum $= 1$, product $= -15$
   D. sum $= -15$, product $= -1$
   E. sum $= 15$, product $= 1$

LEVEL 4

$$-2x^2 + bx + 5$$

6. In the xy-plane, the graph of the equation above assumes its maximum value at $x = 2$. What is the value of b ?

   F. -8
   G. -4
   H. 4
   J. 8
   K. 10

7. Which of the following best describes the graph of the function $y = x^2 + 4x + 8 = 0$?

   A. an upward facing parabola that intersects the x-axis twice
   B. an upward facing parabola that intersects the x-axis once
   C. an upward facing parabola that does not intersect the x-axis
   D. a downward facing parabola that intersects the x-axis once
   E. a downward facing parabola that does not intersect the x-axis

LEVEL 5

8. Which of the following is a quadratic equation that has $-\frac{5}{7}$ as its only solution?

   F. $49x^2 - 70x + 25 = 0$
   G. $49x^2 + 70x + 25 = 0$
   H. $49x^2 + 35x + 25 = 0$
   J. $49x^2 + 25 = 0$
   K. $49x^2 - 25 = 0$

9. Which of the following quadratic equations has solutions $x = -3u$ and $x = 7v$?

   A. $x^2 - 21uv = 0$
   B. $x^2 - x(7v - 3u) - 21uv = 0$
   C. $x^2 - x(7v - 3u) + 21uv = 0$
   D. $x^2 + x(7v - 3u) - 21uv = 0$
   E. $x^2 + x(7v - 3u) + 21uv = 0$

10. You are given the following system of equations.

    $$dx + ey = f$$
    $$y = x^2$$

    where d, e, and f are integers. For which of the following will there be more than one (x, y) solution, with real-number coordinates for the system?

    F. $e^2 + 4df < 0$
    G. $e^2 - 4df < 0$
    H. $d^2 + 4ef < 0$
    J. $e^2 - 4df > 0$
    K. $d^2 + 4ef > 0$

95

LESSON 43 – PROBLEM SOLVING AND DATA
ADVANCED PROBABILITY

1. A committee consisting of 18 people has decided to choose a team leader. The team leader, who will be chosen at random, CANNOT be any of the 4 committee members that already have leadership roles on other committees. What is the probability that Jessica, who does NOT have a leadership role on another committee, will be chosen as the team leader for this committee?

   A. 0
   B. $\frac{1}{22}$
   C. $\frac{1}{18}$
   D. $\frac{1}{14}$
   E. 1

LEVEL 3

2. There is a 30% chance that it will rain on Monday and a 25% chance that it will rain on Tuesday. Assuming that the two events are independent, what is the probability that it will rain both days?

   F. 0.07
   G. 0.075
   H. 0.275
   J. 0.55
   K. 0.75

3. Let $p(E)$ denote the probability that event E occurs, and let $p(\sim E)$ denote the probability that event E does not occur. Which of the following statements is **always** true?

   A. $p(E) + p(\sim E) = 1$
   B. $p(E) < p(\sim E)$
   C. $p(E) > p(\sim E)$
   D. $p(E) > 1$
   E. $p(\sim E) > 1$

4. Set X contains only the integers 0 through 180 inclusive. If a number is selected at random from X, what is the probability that the number selected will be greater than 114 ?

   F. $\frac{65}{181}$
   G. $\frac{66}{181}$
   H. $\frac{11}{30}$
   J. $\frac{66}{179}$
   K. $\frac{67}{180}$

$$A, I, T, R, P$$

5. If the letters from the list above are to be randomly ordered, what is the probability that the letters will appear in the order T, A, P, I, R ?

   A. $\frac{1}{120}$
   B. $\frac{1}{60}$
   C. $\frac{1}{40}$
   D. $\frac{1}{30}$
   E. $\frac{1}{25}$

LEVEL 4

6. A set of numbers consists of the even integers that are greater than 50 and less than 100. What is the probability that a number picked at random from the set will be divisible by 8 ?

   F. $\frac{1}{4}$

   G. $\frac{1}{3}$

   H. $\frac{7}{12}$

   J. $\frac{3}{8}$

   K. $\frac{3}{4}$

7. Maria has 6 dresses and 6 scarves, and each dress matches a different scarf. If she chooses one of these dresses and one of these scarves at random, what is the probability that they will NOT match?

   A. $\frac{1}{36}$

   B. $\frac{1}{6}$

   C. $\frac{2}{3}$

   D. $\frac{5}{6}$

   E. $\frac{35}{36}$

LEVEL 5

8. The integers 1 through 5 are written on each of five cards. The cards are shuffled and one card is drawn at random. That card is then replaced, the cards are shuffled again and another card is drawn at random. This procedure is repeated one more time (for a total of three times). What is the probability that the sum of the numbers on the three cards drawn was between 13 and 15, inclusive?

   F. 0.04
   G. 0.05
   H. 0.06
   J. 0.07
   K. 0.08

9. Joe and Dave are playing a game where two 6-sided dice are rolled. Joe will be awarded 3 points for each of the two dice that shows a six as the top face. Let the random variable x represent the total number of points that Joe receives on any toss of the dice. What is the expected value of x ?

   A. 0
   B. 1
   C. 2.5
   D. 5
   E. 10

10. Suppose that x will be randomly selected from the set $\{-\frac{3}{2}, -1, -\frac{1}{2}, 0, \frac{5}{2}\}$ and that y will be randomly selected from the set $\{-\frac{3}{4}, -\frac{1}{4}, 2, \frac{11}{4}\}$. What is the probability that $\frac{x}{y} < 0$?

   F. $\frac{1}{100}$

   G. $\frac{1}{20}$

   H. $\frac{3}{20}$

   J. $\frac{1}{3}$

   K. $\frac{2}{5}$

97

LESSON 44 – GEOMETRY
ADVANCED TRIGONOMETRY

LEVEL 3

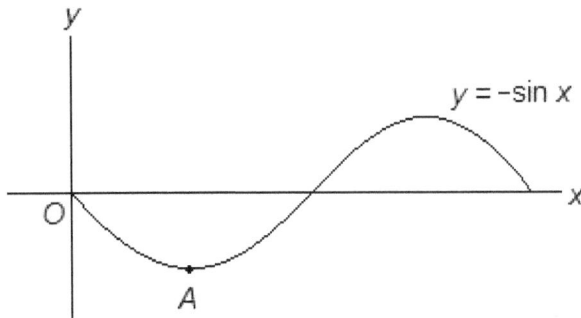

1. The figure above shows one cycle of the graph of the function $y = -\sin x$ for $0 \le x \le 2\pi$. If the minimum value of the function occurs at point A, then the coordinates of A are

   A. $\left(\frac{\pi}{3}, -\pi\right)$

   B. $\left(\frac{\pi}{3}, -1\right)$

   C. $\left(\frac{\pi}{3}, 0\right)$

   D. $\left(\frac{\pi}{2}, -\pi\right)$

   E. $\left(\frac{\pi}{2}, -1\right)$

2. A 17 foot ladder rests against the side of a wall and reaches a point that is 11 feet above the ground. Which of the following expressions is closest to the angle of inclination between the bottom of the ladder and the horizontal floor?

   F. $\sin^{-1}\frac{11}{17}$

   G. $\sin^{-1}\frac{17}{11}$

   H. $\cos^{-1}\frac{11}{17}$

   J. $\cos^{-1}\frac{17}{11}$

   K. $\tan^{-1}\frac{11}{17}$

LEVEL 4

3. In $\triangle ABC$ shown below, the measure of $\angle B$ is 70°, the measure of $\angle C$ is 60°, and $\overline{BC}$ is 10 inches long. Which of the following is an expression for the length, in inches of $\overline{AB}$?

   A. $\frac{10 \sin 50°}{\sin 70°}$

   B. $\frac{10 \sin 60°}{\sin 50°}$

   C. $\frac{10 \sin 130°}{\sin 70°}$

   D. $\frac{10 \sin 70°}{\sin 50°}$

   E. $\frac{10 \sin 60°}{\sin 70°}$

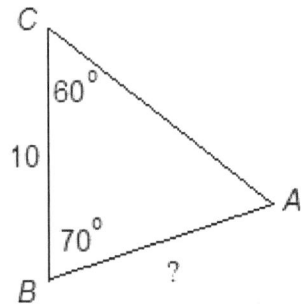

4. In $\triangle ABC$ below, $AB = 12$ inches. To the nearest tenth of an inch, $BC = $?

   F. 9.8

   G. 10.6

   H. 13.5

   J. 13.6

   K. 13.9

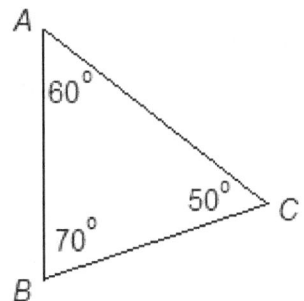

5. If $0 \leq x \leq 2\pi$, $\tan x < 0$ and $\cos x \tan x > 0$, then which of the following is a possible value for x ?

   A. $\frac{\pi}{6}$

   B. $\frac{\pi}{2}$

   C. $\frac{5\pi}{6}$

   D. $\frac{7\pi}{6}$

   E. $\frac{11\pi}{6}$

LEVEL 5

6. Points P and Q lie on a circle of radius 8 with center O. If the measure of $\angle OPQ$ is 50°, what is the length of chord $\overline{PQ}$ to the nearest tenth?

   F. 10.0
   G. 10.1
   H. 10.2
   J. 10.3
   K. 10.4

7. Triangle PQR is shown in the figure below. The measure of $\angle P$ is 32°, $PQ = 9$ in, and $PR = 15$ in. Which of the following is the length, in inches, of $\overline{QR}$?

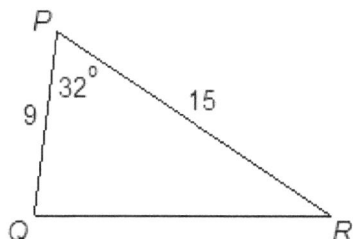

   A. $9 \sin 32°$
   B. $15 \sin 32°$
   C. $\sqrt{15^2 - 9^2}$
   D. $\sqrt{15^2 + 9^2}$
   E. $\sqrt{15^2 + 9^2 - 2(15)(9)\cos 32°}$

8. Which of the following polar coordinates represents the same location as $(5, 30°)$?

   F. $(5, -330°)$
   G. $(5, -300°)$
   H. $(5, -30°)$
   J. $(5, 150°)$
   K. $(5, 330°)$

9. A ladder rests against the side of a wall and reaches a point that is h meters above the ground. The angle formed by the ladder and the ground is $\theta°$. A point on the ladder is k meters from the wall. What is the vertical distance, in meters, from this point on the ladder to the ground?

   A. $(h - k) \tan \theta°$
   B. $(h - k) \cos \theta°$
   C. $h - k \sin \theta°$
   D. $h - k \cos \theta°$
   E. $h - k \tan \theta°$

10. It is given that $\cos x = k$, where x is the radian measure of an angle and $\pi < x < \frac{3\pi}{2}$. If $\cos z = -k$, which of the following could not be the value of z ?

    F. $x - \pi$
    G. $\pi - x$
    H. $2\pi - x$
    J. $3\pi - x$
    K. $x - 5\pi$

99

LESSON 45 – NUMBER THEORY
ADDITIONAL PRACTICE 3

LEVEL 1

1. The absolute value of which of the following numbers is the greatest?

   A. -0.7
   B. -0.073
   C. -0.0079
   D. 0.07
   E. 0.078

2. The first term is -2 in the geometric sequence $-2, 4, -8,\dots$. What is the EIGHTH term of the geometric sequence?

   F. -512
   G. -256
   H. 128
   J. 256
   K. 512

LEVEL 2

3. What is the least common denominator of $\frac{1}{9}, \frac{5}{6}$, and $\frac{7}{10}$?

   A. 15
   B. 90
   C. 135
   D. 270
   E. 540

4. When we subtract $2 - 3i$ from $-5 + 6i$ we get which of the following complex numbers?

   F. $-10 - 18i$
   G. $-7 + 3i$
   H. $-7 + 9i$
   J. $-3 - 3i$
   K. $-3 + 3i$

LEVEL 3

5. For $i = \sqrt{-1}, \frac{1}{1-i} \cdot \frac{1+i}{1+i} = ?$

   A. $1 + i$
   B. $1 - i$
   C. $i - 1$
   D. $\frac{1+i}{2}$
   E. $\frac{1-i}{2}$

LEVEL 4

6. There are 6 red, 6 brown, 6 yellow, and 6 gray scarves packaged in 24 identical, unmarked boxes, 1 scarf per box. What is the least number of boxes that must be selected in order to be sure that among the boxes selected 3 or more contain scarves of the same color?

   F. 6
   G. 7
   H. 8
   J. 9
   K. 10

7. Cards numbered from 1 through 2012 are distributed, one at a time, into nine stacks. 0The card numbered 1 is placed on stack 1, card number 2 on stack 2, card number 3 on stack 3, and so on until each stack has one card. If this pattern is repeated, the card numbered 2012 will be placed on the nth stack. What is the value of n?

    A. 1
    B. 2
    C. 3
    D. 4
    E. 5

8. If $i^2 = -1$, then $\frac{(3+5i)(3-5i)}{17} + i^{86} =$

    F. -3
    G. -2
    H. -1
    J. 0
    K. 1

9. What is the set of all values of k that satisfy the equation $(x^5)^{k^2-9} = 1$ for all nonzero values of x ?

    A. $\{0\}$
    B. $\{3\}$
    C. $\{9\}$
    D. $\{-2,2\}$
    E. $\{-3,3\}$

10. What is the value of $\log_4 64$?

    F. 3
    G. 4
    H. 6
    J. 10
    K. 16

LEVEL 5

11. Consider the fractions $\frac{1}{p}, \frac{1}{q}, \frac{1}{r}$, and $\frac{1}{t}$, where p and q are distinct prime numbers greater than 7, $r = 5p$, and $t = 7q$. Suppose that $p \cdot q \cdot r \cdot t$ is used as the common denominator when finding the sum of these fractions. In order for the sum to be in lowest terms, its numerator and denominator must be reduced by a factor of which of the following?

    A. 35
    B. pq
    C. rt
    D. $pqrt$
    E. $35pqrt$

12. 125 blank cards are lined up in one long row. In the upper left-hand corner of each card a number is written beginning with 1 on the first card, 2 on the second card, and so on until 125 is written in the upper left-hand corner of the last card in the row. Now another number is written in the lower right-hand corner of each card, this time beginning with 125 on the first card, 124 on the second card, and so on until 1 is written in the lower right-hand corner of the last card. Which of the following is a pair of numbers written on the same card?

    F. 71 and 57
    G. 70 and 56
    H. 69 and 55
    J. 68 and 54
    K. 67 and 53

LESSON 46 – ALGEBRA
ADDITIONAL PRACTICE 3

LEVEL 1

1. If $\sqrt{h} = k$ and $k = 25$, then $h = $?

   A.   5
   B.   12.5
   C.   50
   D.   250
   E.   625

LEVEL 2

2. If $\frac{7}{x} = 0.7$, then $x = $?

   F.   0.07
   G.   0.1
   H.   0.7
   J.   7
   K.   10

3. Let a function of 3 variables be defined by $f(x, y, z) = xy^2z + 3yz^2 - 5z$. What is the value of $f(4, -1, 1)$?

   A.   −4
   B.   2
   C.   6
   D.   9
   E.   12

4. If $|3c + 2| < 11$, which of the following cannot be equal to c ?

   F. −4
   G. 0
   H. 1
   J. 2
   K. 3

LEVEL 3

5. When 7 is increased by $5x$, the result is less than 62. What is the greatest possible integer value for x ?

   A.   5
   B.   8
   C.   10
   D.   12
   E.   14

6. The system of equations below has one solution (a, b). What is the value of b ?

   $$x + y = 1$$
   $$2x + y = 3$$

   F.   −2
   G.   −1
   H.   0
   J.   3
   K.   7

102

7. Which of the expressions below is a factor of the polynomial $2x^3 - 9x^2 - 5x$?

    I. x
    II. $x - 5$
    III. $2x + 1$

  A. I only
  B. II only
  C. I and II only
  D. II and III only
  E. I, II, and III

8. What is the determinant of the matrix $\begin{bmatrix} 5 & -6 \\ 3 & -2 \end{bmatrix}$?

  F. -36
  G. -28
  H. 8
  J. 27
  K. 36

LEVEL 4

9. The functions f and g are defined as $f(x) = 2x - 5$ and $g(x) = 3x + 1$. Which of the following expressions is equivalent to $f(g(x))$?

  A. $5x - 4$
  B. $6x - 3$
  C. $6x - 14$
  D. $6x^2 - 5$
  E. $6x^2 - 9x - 5$

10. A portion of the graph of the function g is shown in the xy-plane below. What is the x-intercept of the graph of the function h defined by $h(x) = g(x - 1)$?

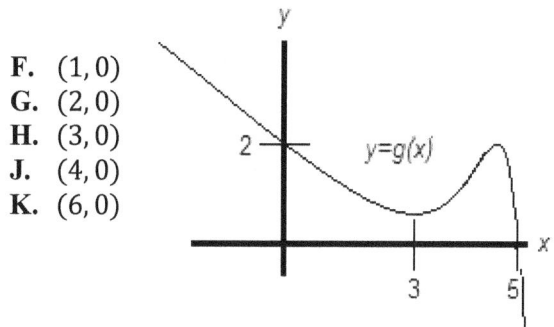

  F. $(1, 0)$
  G. $(2, 0)$
  H. $(3, 0)$
  J. $(4, 0)$
  K. $(6, 0)$

11. If $x = 5$ is a solution to the equation $x^2 - 3x + a = 0$, then the other solution is

  A. -3
  B. -2
  C. -1
  D. 0
  E. 1

LEVEL 5

12. The equation $y = \dfrac{3x^2 - 27}{x^2 - 2x}$ has 2 vertical asymptotes and 1 horizontal asymptote. What are the equations of these 3 asymptotes?

  F. $x = 0, x = 2, y = 3$
  G. $x = 0, x = 2, y = \dfrac{27}{2}$
  H. $x = 2, y = -3, y = 3$
  J. $x = 3, y = 0, y = 2$
  K. $x = \dfrac{27}{2}, y = 0, y = 2$

LESSON 47 – PROBLEM SOLVING AND DATA
ADDITIONAL PRACTICE 3

LEVEL 1

1. Jerry's math class has a homework list of 4 algebra problems, 3 probability problems, and 2 geometry problems. Jerry will select one algebra problem, one probability problem, and one geometry problem from the list to complete the homework assignment. How many different choices of an algebra problem, a probability problem, and a geometry problem are possible?

    A. 3
    B. 9
    C. 18
    D. 24
    E. 48

2. In a jar, there are exactly 100 marbles, each of which is yellow, purple, or blue. The probability of randomly selecting a yellow marble from the jar is $\frac{3}{20}$ and the probability of randomly selecting a purple marble from the jar is $\frac{11}{20}$. How many marbles in the jar are blue?

    F. 25
    G. 30
    H. 32
    J. 35
    K. 40

LEVEL 2

3. If Edna drove s miles in t hours, which of the following represents her average speed, in miles per hour?

    A. $\frac{s}{t}$
    B. $\frac{t}{s}$
    C. $\frac{1}{st}$
    D. st
    E. $s^2 t$

4. What is the median of the list of the numbers below?

    $$27, 7, 15, 8, 4, 17, 2, 5, 8, 2, 5$$

    F. 4
    G. 5
    H. 6
    J. 7
    K. 8

LEVEL 3

5. The average of $x, y, z,$ and w is 12 and the average of z and w is 7. What is the average of x and y ?

    A. 48
    B. 38
    C. 34
    D. 20
    E. 17

6. There are y bricks in a row. If one brick is to be selected at random, the probability that it will NOT be cracked is $\frac{6}{7}$. In terms of y, how many of the bricks are cracked?

   F. $\frac{y}{7}$

   G. $\frac{5y}{7}$

   H. $\frac{7y}{5}$

   J. $\frac{12y}{7}$

   K. $7y$

7. A scientist is testing 5 different liquids. For each test, the scientist chooses 3 of the liquids and mixes them together. What is the least number of tests that must be done so that every possible combination of liquids is tested?

   A. 5
   B. 10
   C. 60
   D. 84
   E. 125

LEVEL 4

8. Joseph drove from home to work at an average speed of 30 miles per hour and returned home along the same route at an average speed of 45 miles per hour. If his total driving time for the trip was 3 hours, how many <u>minutes</u> did it take Joseph to drive from work to home?

   F. 135
   G. 72
   H. 60
   J. 50
   K. 30

List A: 7, a, b, c
List B: 2, 3, 6, a, b, c

9. If the average (arithmetic mean) of the 4 numbers in list A is 10, what is the average of the 6 numbers in list B ?

   A. 44
   B. 22
   C. 11
   D. $\frac{22}{3}$
   E. $\frac{11}{3}$

10. 1000 pet owners were polled to determine which of the following animals they had as pets: cats (C), dogs (D), or birds (B). The Venn diagram below shows the results of the poll except that two of the numbers are missing. If the total number of pet owners polled that said they had dogs as pets is equal to the total number of pet owners polled that said they had birds as pets, how many of the pet owners polled said they have cats as pets?

   F. 169
   G. 335
   H. 380
   J. 501
   K. 513

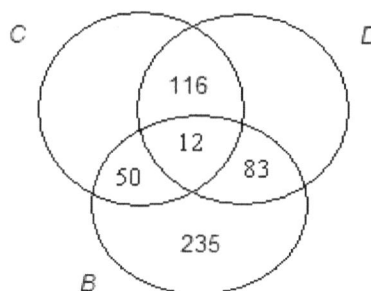

105

LEVEL 5

11. A five-digit number is to be formed using each of the digits 1, 2, 3, 4, and 5 exactly once. How many such numbers are there in which the digits 3 and 4 are not next to each other?

    A. 12
    B. 18
    C. 36
    D. 60
    E. 72

12. A jar contains 1 black marble, 3 white marbles, and 4 yellow marbles. A marble is drawn at random and returned to the jar, then a second marble is drawn at random. What is the probability that the first marble is yellow and the second marble is black?

    F. $\frac{1}{16}$
    G. $\frac{1}{5}$
    H. $\frac{1}{4}$
    J. $\frac{3}{4}$
    K. $\frac{5}{6}$

LESSON 48 – GEOMETRY
ADDITIONAL PRACTICE 3

LEVEL 1

1. In the standard (x, y) coordinate plane, point M with coordinates $(3, 7)$ is the midpoint of $\overline{PQ}$, and P has coordinates $(1, 9)$. What are the coordinates of Q ?

   A. $(5, 5)$
   B. $(-5, -5)$
   C. $(-1, 11)$
   D. $(4, 16)$
   E. $(16, 4)$

2. In the right triangle below, what is the value of y ?

   F. 15
   G. 18
   H. 21
   J. 30
   K. 60

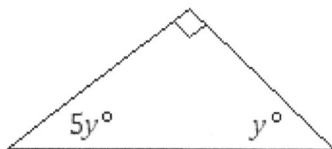

LEVEL 2

3. Right triangle ΔPQR is shown below. The side lengths are given in inches. What is $\cos R$?

   A. $\frac{5}{13}$
   B. $\frac{5}{12}$
   C. $\frac{12}{13}$
   D. $\frac{13}{12}$
   E. $\frac{13}{5}$

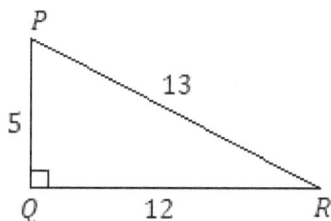

LEVEL 3

4. A chord 48 meters long is 10 meters from the center of a circle, as shown below. What is the diameter of the circle?

   F. 13
   G. 17
   H. 26
   J. 39
   K. 52

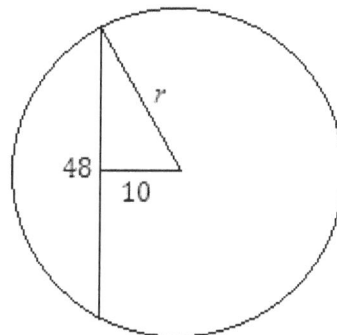

5. In parallelogram $CRAB$, which of the following must be true about the measures of $\angle CRA$ and $\angle RAB$?

   A. each are 90°
   B. each are less than 90°
   C. each are greater than 90°
   D. they add up to 90°
   E. they add up to 180°

6. In a right triangle, one angle measures $x°$, where $\cos x° = \frac{2}{3}$. What is $\sin((90 - x)°)$?

   F. $\frac{2}{3}$
   G. $\frac{\sqrt{5}}{3}$
   H. $\frac{2\sqrt{5}}{5}$
   J. $\frac{3\sqrt{5}}{5}$
   K. $\frac{\sqrt{5}}{3}$

107

LEVEL 4

7. If a 2-centimeter cube were cut in half in all three directions, then in square centimeters, the total surface area of the separated smaller cubes would be how much greater than the surface area of the original 2-centimeter cube?

   A.   0
   B.   12
   C.   24
   D.   36
   E.   48

8. Which of the following is the equation of a line in the xy-plane that is perpendicular to the line with equation $y = 3$?

   F. $y = -3$
   G. $y = -\frac{1}{3}$
   H. $x = -2$
   J. $y = -3x$
   K. $y = -\frac{1}{3}x$

9. The vertex of $\angle P$ is the origin of the standard (x, y) coordinate plane. One ray of $\angle P$ is the positive x-axis. The other ray, $\overrightarrow{PQ}$, is positioned so that $\tan P < 0$ and $\sin P > 0$. In which quadrant, if it can be determined, is point Q ?

   A.   Quadrant I
   B.   Quadrant II
   C.   Quadrant III
   D.   Quadrant IV
   E.   Cannot be determined from the given information

LEVEL 5

10. In the figure below, what is the average of x, y, z and w in terms of k ?

    F. $\frac{k}{4}$
    G. $\frac{k}{2}$
    H. k
    J. $2k$
    K. $4k$

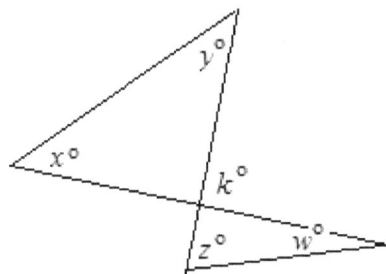

11. The circumference of the base of a right circular cone is 10π and the circumference of a parallel cross section is 8π. If the distance between the base and the cross section is 6, what is the height of the cone?

    A.   30
    B.   20
    C.   15
    D.   7.5
    E.   4.8

12. In the right triangle below, $b > a > 0$. One of the angle measures in the triangle is $\cos^{-1}\frac{b}{\sqrt{a^2+b^2}}$. What is $\csc[\cos^{-1}\left(\frac{b}{\sqrt{a^2+b^2}}\right)]$?

    F. $\frac{b}{a}$
    G. $\frac{a}{b}$
    H. $\frac{b}{\sqrt{a^2+b^2}}$
    J. $\frac{a}{\sqrt{a^2+b^2}}$
    K. $\frac{\sqrt{a^2+b^2}}{a}$

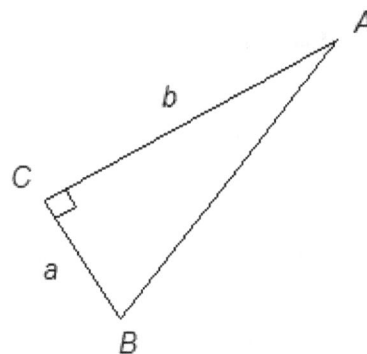

108

PROBLEMS BY LEVEL AND TOPIC
PROBLEM SET A

Full solutions to these problems are available for free download here:
www.SATPrepGet800.com/UniACTyWB

LEVEL 1: NUMBER THEORY

1. $|5(-4) + 3(5)| = ?$

   A. -35
   B. -5
   C. 5
   D. 35
   E. 36

2. Which of the following inequalities orders the 0.1, 0.02, and $\frac{1}{8}$ from greatest to least?

   F. $0.1 > 0.02 > \frac{1}{8}$
   G. $0.02 > 0.1 > \frac{1}{8}$
   H. $0.02 > \frac{1}{8} > 0.1$
   J. $\frac{1}{8} > 0.02 > 0.1$
   K. $\frac{1}{8} > 0.1 > 0.02$

3. What is the greatest integer less than $\sqrt{79}$?

   A. 7
   B. 8
   C. 9
   D. 10
   E. 11

4. Which of the following lists all the positive factors of 27 ?

   F. $1, 27$
   G. $3, 9$
   H. $3, 9, 27$
   J. $27, 54, 81$
   K. $1, 3, 9, 27$

109

5. What is the least common multiple of 3, 7 and 14 ?

    A. 168
    B. 126
    C. 84
    D. 42
    E. 28

6. The first term is 2 in the geometric sequence 2, 4, 8, 16,…. What is the EIGHTH term of the geometric sequence?

    F. 32
    G. 64
    H. 128
    J. 256
    K. 512

7. If $(-1 + i) + (-1 - 2i) = a + bi$ and $i = \sqrt{-1}$, then what is the value of ab ?

    A. 6
    B. 5
    C. 4
    D. 3
    E. 2

8. For $i = \sqrt{-1}$, the sum $(3 - 7i) + (5 + 4i)$ is

    F. $2 + 11i$
    G. $2 - 3i$
    H. $-2 - 11i$
    J. $8 - 3i$
    K. $8 - 11i$

LEVEL 1: ALGEBRA

9. For what value of x is the equation $x + 5(x - 4) = 4$ true?

    A. 24
    B. 20
    C. 8
    D. 4
    E. 3

10. The population of rabbits on Rabbit Island is modeled by the equation $P = 1000(1.2)^t$, where t is the number of years after January 1, 2017. Based on the model, which of the following numbers is closest to the population of rabbits on Rabbit Island on January 1, 2020 ?

    F. 1200
    G. 1400
    H. 1700
    J. 2000
    K. 2500

11. The expression $x[y + (z - w)]$ is equivalent to

    A. $xy + z - w$
    B. $xy + z + w$
    C. $xy + xz - w$
    D. $xy + xz + xw$
    E. $xy + xz - xw$

12. Which of the following is a value for z that solves the equation $|z - 4| = 9$?

    F. -13
    G. -5
    H. $\frac{9}{4}$
    J. 5
    K. 36

13. If $a = -5$, what is the value of $\frac{a^2-4}{a+2}$?

    A. -7
    B. -4
    C. 4
    D. $9\frac{2}{3}$
    E. 12

14. When written in symbols, "The square of the sum of a and b" is represented as:

    F. $(a + b)^2$
    G. $a^2 + b$
    H. $a + b^2$
    J. $a^2 + b^2$
    K. $(a^2 + b^2)^2$

15. If $x + 3 = 10$, then $(x + 1)^2 =$

    A. 16
    B. 25
    C. 36
    D. 49
    E. 64

16. If $A = -5x$ and $B = 3y - x$, then what is the value of $A - B$?

    F. $-6x - 3y$
    G. $-6x + 3y$
    H. $-4x - 3y$
    J. $-4x + 3y$
    K. $\ \ \ 4x - 3y$

LEVEL 1: PROBLEM SOLVING AND DATA

17. If a 6-pound quiche is cut into three equal pieces and each of those pieces is cut into four equal pieces, what is the weight, in ounces, of each piece of quiche? (1 pound = 16 ounces)

    A. 2
    B. 4
    C. 6
    D. 8
    E. 10

18. Joseph bought a tie for 60% of its original price of \$14.50 and a shirt for $\frac{2}{5}$ of the original price of \$45.00. Ignoring sales tax, what is the total amount of these purchases?

    F. \$21.00
    G. \$25.00
    H. \$26.00
    J. \$26.70
    K. \$40.50

19. To keep up with inflation, a store owner raises the price of a \$30 item by 26%. What is the new price of the item?

    A. \$30.22
    B. \$32.20
    C. \$37.00
    D. \$37.80
    E. \$52.00

20. The average (arithmetic mean) of five numbers is 80. If three of the numbers are 32, 62 and 82, what is the sum of the other two?

    F. 224
    G. 112
    H. 96
    J. −19.2
    K. −96

21. In a math class, a student's overall grade for the semester is determined by throwing out the lowest test grade and taking the average of the remaining test grades. Alena received test grades of 67, 71, 83, 91, 95, and 98 this semester. What overall grade did Alena receive in the math class this semester?

    A. 82
    B. 84.2
    C. 87
    D. 87.6
    E. 92

22. A menu lists 4 meals, 6 drinks, and 3 desserts. How many different ways are there to choose one meal, one drink, and one dessert from this menu?

    F. 13
    G. 18
    H. 22
    J. 27
    K. 72

23. Markus has 4 white hats and 5 black hats in his closet. If he randomly takes 1 of these 9 hats from his closet, what is the probability that the hat that Markus takes is black?

    A. $\dfrac{1}{9}$

    B. $\dfrac{1}{5}$

    C. $\dfrac{4}{9}$

    D. $\dfrac{5}{9}$

    E. $\dfrac{5}{4}$

24. If the probability that it will be sunny tomorrow is 0.4, what is the probability that it will <u>not</u> be sunny tomorrow?

    F. 1.4
    G. 1.0
    H. 0.6
    J. 0.1
    K. 0.0

LEVEL 1: GEOMETRY

25. In the figure below, three lines intersect at a point. What is the value of k ?

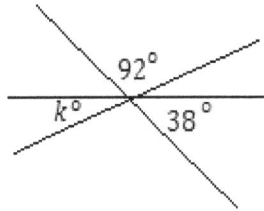

A. 38
B. 50
C. 54
D. 92
E. 130

26. A point at $(6, -5)$ in the standard (x, y) coordinate plane is reflected in the x-axis. What are the coordinates of the new point?

F. $(-6, -5)$
G. $(-6, 5)$
H. $(6, 5)$
J. $(-5, 6)$
K. $(5, -6)$

27. If the degree measure of one of the three angles of a triangle is 80° and the other two angles are congruent, what is the measure of one of the congruent angles?

A. 40°
B. 50°
C. 80°
D. 100°
E. 120°

28. In the figure below, A, B, and C lie on the same line. B is the center of the smaller circle, and C is the center of the larger circle. If the radius of the smaller circle is 7, what is the diameter of the larger circle?

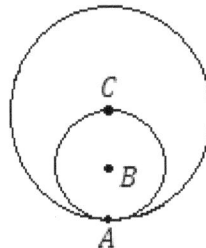

F. 10.5
G. 12
H. 14
J. 21
K. 28

29. The interior dimensions of a cube are 5 inches by 5 inches by 5 inches. What is the volume, in cubic inches, of the interior of the cube?

A. 15
B. 30
C. 120
D. 125
E. 150

114

30. What is the area, in feet, of a rectangle with length 3 ft and width 15 ft?

    F. 18
    G. 21
    H. 36
    J. 45
    K. 90

31. If the perimeter of the rectangle below is 78, what is the value of x ?

    A. 20
    B. 19
    C. 18
    D. 17
    E. 16

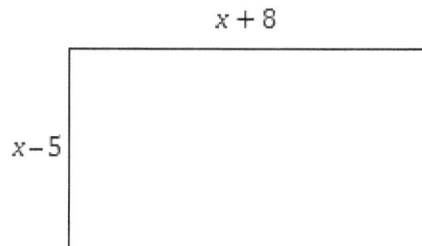

$x+8$

$x-5$

32. What is the slope of the line through $(-2,-1)$ and $(-5,3)$ in the standard (x,y) coordinate plane?

    F. $-\frac{4}{3}$
    G. $-\frac{3}{4}$
    H. 0
    J. $\frac{3}{4}$
    K. $\frac{4}{3}$

LEVEL 2: NUMBER THEORY

33. If m is an odd integer, what is the greatest even integer less than m ?

    A. $m-3$
    B. $m-2$
    C. $m-1$
    D. $2(m-1)$
    E. $2(m-1)-3$

34. If you add up 5 consecutive odd integers that are each less than 32, what is the largest possible sum?

    F. 25
    G. 87
    H. 125
    J. 135
    K. 145

35. The expression $\dfrac{5+\frac{1}{5}}{2+\frac{1}{10}}$ is equal to:

    A. $\dfrac{7}{2}$

    B. $\dfrac{52}{21}$

    C. 3

    D. 7

    E. 9

36. If $c > 1$, then which of the following has the <u>least</u> value?

    F. $\sqrt{c}$
    G. $\sqrt{2c}$
    H. $\sqrt{c^2}$
    J. $c\sqrt{c}$
    K. c^2

37. What is the least common denominator of the fractions $\dfrac{1}{4}$, $\dfrac{5}{6}$, and $\dfrac{11}{15}$?

    A. 15
    B. 60
    C. 90
    D. 180
    E. 360

$$\frac{6}{n}, \frac{7}{n}, \frac{11}{n}$$

38. If each of the fractions above is in its simplest reduced form, then which of the following could be the value of n ?

    F. 15
    G. 25
    H. 27
    J. 35
    K. 55

39. The first term in the geometric sequence below is -1. If it can be determined, what is the 5th term of the sequence?

$$-1, 3, -9, \ldots$$

    A. -243
    B. -81
    C. 81
    D. 243
    E. Cannot be determined from the given information

40. If $2^{x+1} - 5 = 59$, what is the value of x ?

    F. 6
    G. 5
    H. 4
    J. 3
    K. 2

LEVEL 2: ALGEBRA

41. For all x, $7 - 5(2 - x) = $?

    A. $-5x + 17$
    B. $-5x - 1$
    C. $-5x - 3$
    D. $5x - 1$
    E. $5x - 3$

42. A function f is defined as $f(x, y, z) = xy^2 - x^2z + yz$. What is $f(-1, -2, 3)$?

    F. -13
    G. -7
    H. -5
    J. -1
    K. 1

43. What is the sum of the solutions of the 2 equations below?

$$6x = 15$$
$$2y + 3 = 11$$

    A. $1\frac{1}{5}$
    B. $6\frac{1}{2}$
    C. 7
    D. 8
    E. $15\frac{1}{2}$

$$ax - bx + cx - dx$$

44. For all real numbers a, b, c, and d, the expression above can be written as the product of x and which of the following?

    F. $-a + b - c + d$
    G. $-a + b - c - d$
    H. $-a - b - c - d$
    J. $a - b + c - d$
    K. $a + b + c + d$

45. If $|3 - 8x| > 34$, which of the following is a possible value of x ?

    A. -4
    B. -2
    C. 2
    D. 3
    E. 4

46. Given that $\sqrt{3x} - 6 = 3$, $x = ?$

    F. -27
    G. -9
    H. 3
    J. 27
    K. 54

47. Which of the following augmented matrices represents the system of linear equations below?

$$2x + 7y = 3$$
$$x - 3y = -5$$

    A. $\begin{bmatrix} 2 & 7 & | & -3 \\ 1 & -3 & | & 5 \end{bmatrix}$
    B. $\begin{bmatrix} 2 & 7 & | & 3 \\ 1 & -3 & | & -5 \end{bmatrix}$
    C. $\begin{bmatrix} 2 & 0 & | & 3 \\ 1 & -3 & | & -5 \end{bmatrix}$
    D. $\begin{bmatrix} 2 & -7 & | & 3 \\ 1 & -3 & | & -5 \end{bmatrix}$
    E. $\begin{bmatrix} 2 & 1 & | & 3 \\ 7 & -3 & | & -5 \end{bmatrix}$

48. For which nonnegative value of b is the expression $\frac{1}{2-b^2}$ undefined?

    F. 0
    G. 2
    H. 8
    J. $\sqrt{2}$
    K. $\sqrt{8}$

LEVEL 2: PROBLEM SOLVING AND DATA

49. The ratio of 19 to 5 is equal to the ratio of 133 to what number?

    A. 0.714
    B. 35
    C. 102.6
    D. 225
    E. 505.4

50. How many minutes would it take a car to travel 42 miles at a constant speed of 56 miles per hour?

    F. 90
    G. 80
    H. 45
    J. 40
    K. 30

51. * Janice spent 22% of her 7-hour school day in her AP Calculus class. How many <u>minutes</u> of her school day were spent in AP Calculus?

    A. 1.54
    B. 24.3
    C. 46.97
    D. 86.31
    E. 92.4

$$15, 17, 3, 19, 2, 5, 22, 36, b$$

52. If b is the median of the 9 numbers listed above, which of the following could be the value of b ?

    F. 17
    G. 18
    H. 19
    J. 20
    K. 21

53. Four different DVDs are to be stacked in a pile. In how many different orders can the DVDs be placed on the stack?

    A. 10
    B. 16
    C. 24
    D. 27
    E. 256

$$3, \ 6, \ 7, \ 21, \ 27, \ 35, \ 42, \ 63, \ 70$$

54. A number is to be selected at random from the list above. What is the probability that the number selected will be a multiple of both 3 and 7 ?

    F. $\frac{2}{9}$
    G. $\frac{1}{3}$
    H. $\frac{4}{9}$
    J. $\frac{2}{3}$
    K. 1

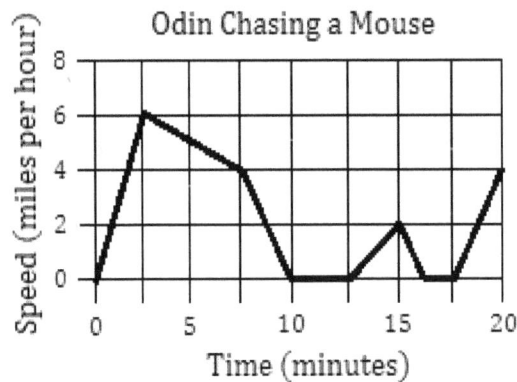

Odin Chasing a Mouse

55. Odin the cat chased a mouse for twenty minutes. His time and speed are displayed in the graph above. According to the graph, which of the following is the best estimate for the number of minutes that Odin was not moving during the chase.

    A. 0
    B. 1
    C. 2
    D. 4
    E. 6

56. A lab technician is testing 4 different liquids. For each test, the technician chooses 2 of the liquids and mixes them together. What is the least number of tests that must be done so that every possible combination of liquids is tested?

    F. 24
    G. 16
    H. 12
    J. 6
    K. 4

LEVEL 2: GEOMETRY

57. The coordinates of the endpoints of $\overline{AB}$, in the standard (x, y) coordinate plane, are $(-3, -5)$ and $(3, 13)$. What is the y-coordinate of the midpoint of $\overline{AB}$?

    A.   0
    B.   2
    C.   4
    D.   9
    E.   10

58. In the figure below, the two triangles are isosceles. If $a + c = 175$ and $a = 22$, what is the value of b ?

   F. 175
   G. 153
   H. 126
   J.   54
   K.   27

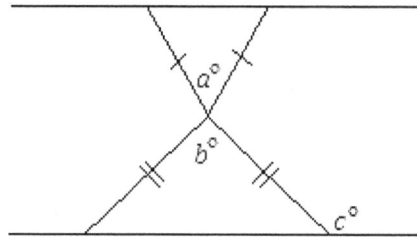

59. In the isosceles right triangle below, $PQ = 7$ inches. What is the length, in inches, of $\overline{PR}$?

   A. $7\sqrt{2}$
   B. $\sqrt{14}$
   C. 14
   D. 7
   E. 3.5

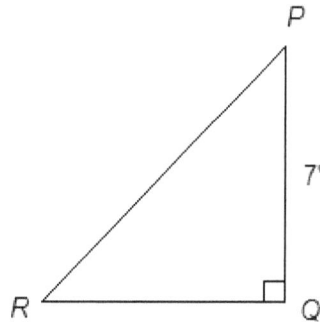

60. In the figure below, point Q lies on side PR. If $48 < y < 50$, which of the following is a possible value of x?

   F. 130
   G. 131
   H. 132
   J. 133
   K. 134

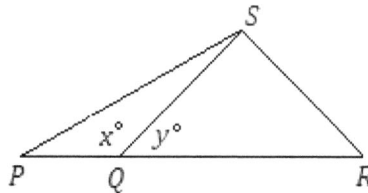

61. The circumference of a circle is 40 cm. What is the length, in centimeters, of the *radius* of the circle?

   A. 2π
   B. $\dfrac{20}{\pi}$
   C. $\dfrac{40}{\pi}$
   D. 40
   E. 40π

121

62. In parallelogram $PQRS$ below, $\overline{PR}$ is a diagonal, the measure of $\angle PSR$ is 120°, and the measure of $\angle PRQ$ is 40°. What is the measure of $\angle SRP$?

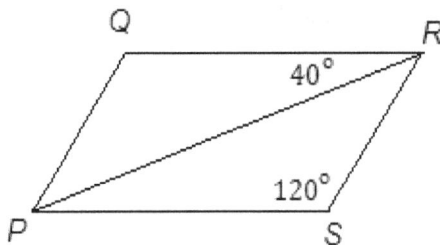

F. 20°
G. 30°
H. 50°
J. 60°
K. 120°

63. Which of the following equations represents the line in the standard (x, y) coordinate plane that passes through $(-6, 22)$ and has a slope of -3 ?

A. $y = -\frac{1}{3}x + 2$
B. $y = -3x + 4$
C. $y = -3x + 2$
D. $y = 3x + 4$
E. $y = \frac{1}{3}x + 2$

64. In ΔPQR below, which of the following trigonometric expressions has value $\frac{5}{12}$?

F. $\tan R$
G. $\tan P$
H. $\sin R$
J. $\sin P$
K. $\cos R$

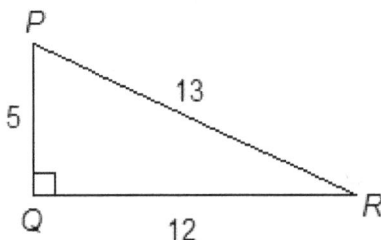

LEVEL 3: NUMBER THEORY

65. In what order should $\frac{7}{3}$, $\frac{9}{4}$, $\frac{11}{5}$, and $\frac{17}{8}$ be listed to be arranged in increasing order?

A. $\frac{7}{3} < \frac{9}{4} < \frac{11}{5} < \frac{17}{8}$
B. $\frac{9}{4} < \frac{11}{5} < \frac{17}{8} < \frac{7}{3}$
C. $\frac{11}{5} < \frac{17}{8} < \frac{9}{4} < \frac{7}{3}$
D. $\frac{17}{8} < \frac{11}{5} < \frac{7}{3} < \frac{9}{4}$
E. $\frac{17}{8} < \frac{11}{5} < \frac{9}{4} < \frac{7}{3}$

66. On Monday Jason spent one third of his allowance. On Tuesday he spent one third of the remaining money, and on Wednesday he spent one third of what remained from Tuesday. If $8 then remained, how much did he originally receive for his allowance?

    F. $12
    G. $18
    H. $27
    J. $54
    K. $96

67. For all positive integers k, what is the greatest common factor of the 2 numbers $150k$ and $700k$?

    A. 10
    B. 50
    C. k
    D. $10k$
    E. $50k$

68. Which of the following is the least common denominator for the expression below?

$$\frac{1}{11^2 \cdot 29} + \frac{1}{11 \cdot 19^2 \cdot 29} + \frac{1}{11 \cdot 29^4}$$

    F. $11 \cdot 19$
    G. $11 \cdot 29$
    H. $11 \cdot 19 \cdot 29$
    J. $11^2 \cdot 19^2 \cdot 29^4$
    K. $11^4 \cdot 19^2 \cdot 29^6$

69. Which of the following complex numbers is equivalent to $(6 - i)(5 + 3i)$? (Note: $i = \sqrt{-1}$)

    A. $30 - 3i$
    B. $30 + 3i$
    C. $11 + 2i$
    D. $33 + 13i$
    E. $33 - 13i$

70. For $i = \sqrt{-1}$, $\frac{1}{1+i} \cdot \frac{1-i}{1-i} = ?$

    F. $1 + i$
    G. $1 - i$
    H. $i - 1$
    J. $\frac{1+i}{2}$
    K. $\frac{1-i}{2}$

71. Which of the following expressions is equivalent to $x^{-\frac{2}{3}}$?

    A. $\dfrac{x^2}{3}$

    B. $\dfrac{3}{x^2}$

    C. $-\sqrt[3]{x^2}$

    D. $\dfrac{1}{\sqrt[3]{x^2}}$

    E. $-\dfrac{1}{\sqrt[3]{x^2}}$

72. If $3^x = 11$, then $3^{2x} =$

    F.   5.5
    G.   22
    H.   33
    J.   121
    K.   1331

LEVEL 3: ALGEBRA

73. What value of k will satisfy the equation $0.3(k + 2100) = k$?

    A.   630
    B.   900
    C.   1200
    D.   1500
    E.   1935

74. The operation ■ is defined as $a\ ■\ b = \dfrac{2b^2 - 8a^2}{b + 2a}$ where a and b are real numbers and $b \neq -2a$. What is the value of $(-2)\ ■\ (-1)$?

    F.   14
    G.   12
    H.   6
    J. -2
    K. -6

75. The graphs of the functions $f(x) = 2x + 2$ and $g(x) = 5 - x$ in the standard (x, y) coordinate plane are lines. If it can be determined, at what point do the graphs intersect?

    A. $(-1, 0)$
    B. $(\ 0, 2)$
    C. $(\ 1, 4)$
    D. $(\ 2, 6)$
    E.   Cannot be determined from the given information

76. The system of equations below has 1 solution (c, d). What is the value of d ?

$$2c + 3d = 12$$
$$3c - d = -4$$

    F. -4
    G. -2
    H. 0
    J. 2
    K. 4

77. Which of the following expressions is a factor of the polynomial $x^2 - x - 110$?

    A. $x - 7$
    B. $x - 8$
    C. $x - 9$
    D. $x - 10$
    E. $x - 11$

78. The expression $x^2 - x - 12$ can be written as the product of two binomial factors with integer coefficients. One of the binomials is $(x + 3)$. Which of the following is the other binomial?

    F. $x^2 - 4$
    G. $x^2 + 4$
    H. $x - 4$
    J. $x + 4$
    K. $x + 5$

79. What is the set of real solutions for $|x|^2 - |x| - 12 = 0$?

    A. $\{4\}$
    B. $\{-3, 4\}$
    C. $\{3, 4\}$
    D. $\{-4, 4\}$
    E. $\{-4, -3, 3, 4\}$

80. A child has set up a rows of dominoes with $(b + c)$ dominoes in each row. Which of the following is an expression for the total number of dominoes that the child has set up?

    F. $a + b + c$
    G. $a \cdot b \cdot c$
    H. $a + b \cdot c$
    J. $a \cdot b + a \cdot c$
    K. $a \cdot b + c$

LEVEL 3: PROBLEM SOLVING AND DATA

81. A copy machine makes 3200 copies per hour. At this rate, in how many <u>minutes</u> can the copy machine produce 800 copies?

    A. 4
    B. 8
    C. 12
    D. 15
    E. 18

82. Jessica is planning to bake a pie using a recipe that requires $5\frac{1}{2}$ tablespoons of cinnamon. She has only $2\frac{3}{4}$ tablespoons left. The amount she has left is what fraction of the amount of cinnamon she needs for the recipe?

    F. $\frac{3}{4}$
    G. $\frac{1}{2}$
    H. $\frac{1}{4}$
    J. $\frac{1}{5}$
    K. $\frac{1}{8}$

83. An integer is decreased by 30% and the resulting number is then increased by 35%. The final number is what percent of the original number?

    A. 90
    B. 92
    C. 94.5
    D. 100
    E. 105

84. Jeff has taken 6 of 10 equally weighted math tests this semester, and he has an average score of exactly 82.0 points. How many points does he need to earn on the 7th test to bring his average score up to exactly 83 points?

    F. 86
    G. 87
    H. 88
    J. 89
    K. 90

85. To decrease the mean of 5 numbers by 3, by how much would the sum of the 5 numbers have to decrease?

    A. 3
    B. 5
    C. 7.5
    D. 12
    E. 15

86. How many integers between 9 and 300 have the tens digit equal to 2, 3, or 4 and the units digit (ones digit) equal to 5 or 6 ?

    F. 5
    G. 6
    H. 8
    J. 18
    K. 23

87. A jar contains 11 orange jellybeans, 7 yellow jellybeans, and 4 red jellybeans. How many additional yellow jellybeans must be added to the jar so that the probability of randomly selecting a yellow jellybean is $\frac{6}{11}$?

    A. 8
    B. 11
    C. 14
    D. 17
    E. 23

88. Seven marbles of different colors are lined up in a row. The red marble is placed at the far left of the row. In how many different orders can the remaining marbles be lined up to the right of the red marble?

    F.   21
    G.   28
    H.   720
    J.   5040
    K. 46,656

LEVEL 3: GEOMETRY

89. Which of the following figures in a plane separates it into half-planes?

    A. A point
    B. An angle
    C. A line segment
    D. A ray
    E. A line

127

90. If a rectangle measures 63 centimeters by 84 centimeters, what is the length, in centimeters, of a diagonal of the rectangle?

    F. 55
    G. 67
    H. 73.5
    J. 105
    K. 147

91. A circle in the standard (x, y) coordinate plane has center $O(-2, -6)$ and passes through the point $P(4, 2)$. What is the length of a <u>diameter</u> of the circle?

    A. $2\sqrt{5}$
    B. $4\sqrt{5}$
    C. 10
    D. 20
    E. 40

92. A tank in the shape of a right circular cylinder is completely filled with water as shown below. If the volume of the tank is 160π cubic yards, what is the <u>diameter</u> of the base of the cylinder, in yards?

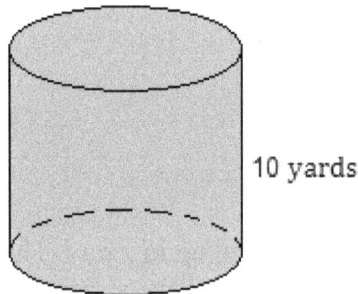

    F. 8
    G. 10
    H. 12
    J. 14
    K. 15

    10 yards

93. If the perimeter of a rectangle is 100 cm then the area of the rectangle, in centimeters, is

    A. 200
    B. 400
    C. 600
    D. 800
    E. Cannot be determined from the given information

94. Which of the following is an equation of the line in the standard (x, y) coordinate plane that passes through the point $(0, -3)$ and is perpendicular to the line $y = 4x + 7$?

    F. $y = -\frac{1}{4}x - 3$
    G. $y = -4x - 3$
    H. $y = 4x - 3$
    J. $y = \frac{1}{4}x - 3$
    K. $y = \frac{1}{4}x + 6$

95. For right triangle ΔPQR below, which of the following expressions has a value that is equal to $\sin P$?

   A. $\sin R$
   B. $\cos P$
   C. $\cos R$
   D. $\tan P$
   E. $\tan R$

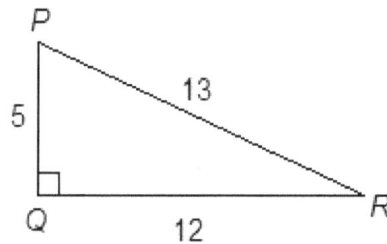

96. In the figure below, line k is parallel to line n. If line m bisects angle ABC, what is the value of x ?

   F. 50
   G. 60
   H. 70
   J. 80
   K. 90

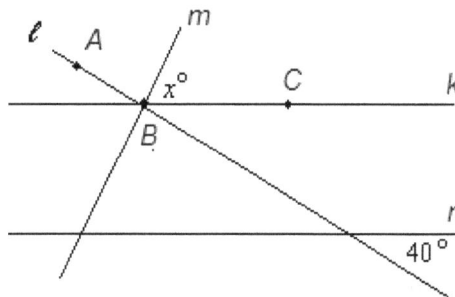

LEVEL 4: NUMBER THEORY

97. If c is a positive odd integer and d is a positive even integer, then $[(+7)(-7)]^{cd}$ is

   A. negative and even
   B. negative and odd
   C. positive and even
   D. positive and odd
   E. zero

98. If $n \leq -5$. which of the following has the least value?

   F. $\dfrac{1}{(n+3)^2}$

   G. $-\dfrac{1}{(n+3)^2}$

   H. $\dfrac{1}{n+3}$

   J. $\dfrac{1}{n+4}$

   K. $\dfrac{1}{n-4}$

99. If b is a positive integer that divides both 98 and 140, but divides neither 20 nor 35, what should you get when you add the digits in b ?

   A. 1
   B. 3
   C. 4
   D. 5
   E. 6

129

100. What is the least positive integer greater than 7 that leaves a remainder of 6 when divided by both 9 and 15 ?

   F. 51
   G. 52
   H. 53
   J. 54
   K. 55

101. Each term of a certain sequence is greater than the term before it. The difference between any two consecutive terms in the sequence is always the same number. If the third and tenth terms of the sequence are 46 and 81, respectively, what is the ninth term?

   A. 72
   B. 74
   C. 75
   D. 76
   E. 78

102. If u and v are real numbers, $i = \sqrt{-1}$, and

$$(u - v) + 3i = 7 + vi,$$

   then what is $u + v$?

   F. 3
   G. 10
   H. 13
   J. 17
   K. 19

103. For how many rational numbers q is the equation $7^{5q+2} = 49^{q-3}$ true?

   A. 0
   B. 1
   C. 2
   D. 3
   E. An infinite number

104. For what real value of y is $\dfrac{2^y 2^9}{(2^4)^8} = \dfrac{1}{32}$ true?

   F. 2
   G. 3
   H. 11
   J. 18
   K. 25

LEVEL 4: ALGEBRA

$$P(x) = \frac{20x}{98 - x}$$

105. The function P above models the monthly profit, in thousands of dollars, for a company that sells x percent of their inventory for the month. If \$90,000 is earned in profit during the month of April, what percent of April's inventory, to the nearest whole percent, has been sold?

    A. 25%
    B. 42%
    C. 56%
    D. 80%
    E. 90%

106. In the standard (x, y) coordinate plane, for what value(s) of x, if any, is there NO value of y such that (x, y) is on the graph of $y = \frac{7-x}{(x-5)(x+4)(x-4)}$?

    F. $-5, -4$, and 4 only
    G. $-4, 4$, and 5 only
    H. -5 only
    J.   5 only
    K. There are no such values of x.

107. The figure below shows the graph of the function g in the xy-plane. Which of the following are true?

    I. $g(b) = 0$
    II. $g(a) + g(b) + g(0) = $ 0
    III. $g(a) > g(b)$

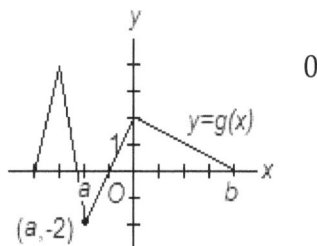

    A. None
    B. I only
    C. II only
    D. I and II only
    E. I, II, and III

108. In the xy-plane, line ℓ is the graph of $5x + ky = 8$, where k is a constant. The graph of $10x + 22y = 17$ is parallel to line ℓ. What is the value of k ?

    F. $\frac{1}{11}$
    G. $\frac{11}{25}$
    H. $\frac{25}{11}$
    J. 11
    K. 25

131

109. If $c > 0$, $s^2 + t^2 = c$, and $st = c + 5$, what is $(s + t)^2$ in terms of c ?

    A. $c + 5$
    B. $c + 10$
    C. $2c + 5$
    D. $2c + 10$
    E. $3c + 10$

110. When a pet store sets the price of cat food at \$1.50 per can, the store sells 300 cans of cat food per week. With each \$0.10 increase in the \$1.50 price per can, the store sells 5 less cans of cat foot per week. Let p be the number of \$0.10 increases in the price per can. Which of the following expressions best represents the dollar amount of the store's weekly sales of cat food?

    F. $(1.60)(300 + 5p)$
    G. $(1.60)(300 - 5p)$
    H. $(300 + 0.10p)(5 + 1.50p)$
    J. $(1.50 + 0.10p)(300 + 5p)$
    K. $(1.50 + 0.10p)(300 - 5p)$

111. What is the matrix product $\begin{bmatrix} -1 \\ 0 \\ 1 \end{bmatrix} \begin{bmatrix} s & t & u \end{bmatrix}$?

    A. $\begin{bmatrix} -s + u \end{bmatrix}$
    B. $\begin{bmatrix} -s & 0 & u \end{bmatrix}$
    C. $\begin{bmatrix} -s \\ 0 \\ u \end{bmatrix}$
    D. $\begin{bmatrix} -s & -t & -u \\ 0 & 0 & 0 \\ s & t & u \end{bmatrix}$
    E. $\begin{bmatrix} -s & 0 & s \\ -t & 0 & t \\ -u & 0 & u \end{bmatrix}$

112. In the equation below, n and k are constants. If the equation is true for all values of x, what is the value of k ?

$$(x - n)(x - 9) = x^2 - 4nx + k$$

    F. 3
    G. 6
    H. 9
    J. 27
    K. 54

LEVEL 4: PROBLEM SOLVING AND DATA

113. For nonzero numbers a, b, and c, if c is three times b and b is $\frac{1}{5}$ of a, what is the ratio of a^2 to c^2 ?

    A. 9 to 25
    B. 25 to 9
    C. 5 to 9
    D. 5 to 3
    E. 3 to 5

114. Twenty-six people were playing a game. 1 person scored 50 points, 3 people scored 60 points, 4 people scored 70 points, 5 people scored 80 points, 6 people scored 90 points, and 7 people scored 100 points. Which of the following correctly shows the order of the median, mode and average (arithmetic mean) of the 26 scores?

    F. average < median< mode
    G. average < mode < median
    H. median < mode < average
    J. median < average < mode
    K. mode < median < average

115. The average (arithmetic mean) of 17 numbers is j. If two of the numbers are k and m, what is the average of the remaining 15 numbers in terms of j, k and m ?

    A. $\dfrac{k+m}{17}$
    B. $17j + k + m$
    C. $\dfrac{16j-k-m}{17}$
    D. $\dfrac{17j-k-m}{15}$
    E. $\dfrac{17(k-m)-j}{15}$

116. A desk is to be painted one color and a chair is to be painted a different color. If 6 different colors are available, how many color combinations are possible?

    F. 6
    G. 11
    H. 12
    J. 15
    K. 30

117. An integer from 37 through 842, inclusive, is to be chosen at random. What is the probability that the number chosen will have 9 as at least one digit?

    A. $\dfrac{76}{403}$

    B. $\dfrac{153}{806}$

    C. $\dfrac{77}{403}$

    D. $\dfrac{5}{26}$

    E. $\dfrac{6}{31}$

118. An urn contains a number of marbles of which 98 are blue, 14 are red, and the remainder are white. If the probability of picking a white marble from this urn at random is $\dfrac{1}{5}$, how many white marbles are in the urn?

    F. 14
    G. 28
    H. 42
    J. 56
    K. 70

119. A pet store has a white dog, a black dog, and a grey dog. The store also has three cats – one white, one black, and one grey – and three birds – one white, one black, and one grey. Jonathon wants to buy one dog, one cat, and one bird. How many different possibilities does he have?

    A. 3
    B. 6
    C. 9
    D. 12
    E. 27

120. Exactly 5 musicians try out to play 5 different instruments for a particular performance. If each musician can play each of the 5 instruments, and each musician is assigned an instrument, what is the probability that Gary will play the piano?

    F. $\dfrac{1}{5}$

    G. $\dfrac{3}{5}$

    H. $\dfrac{1}{25}$

    J. $\dfrac{2}{25}$

    K. $\dfrac{4}{25}$

LEVEL 4: GEOMETRY

121. Three distinct lines, all contained in a plane, separate the plane into distinct regions. Which of the following is a possibility for the number of distinct regions of the plane that may be separated by any 3 such lines?

    A. 2
    B. 3
    C. 5
    D. 6
    E. 8

122. An airplane is flying at a height of 3 miles above the ground. The angle of depression from the airplane to the airport is 60°, as shown in the figure below. What is the distance from the airplane to the airport, in miles?

    F. $\frac{\sqrt{3}}{3}$

    G. $\frac{\sqrt{3}}{2}$

    H. $\sqrt{3}$

    J. 6

    K. $3\sqrt{3}$

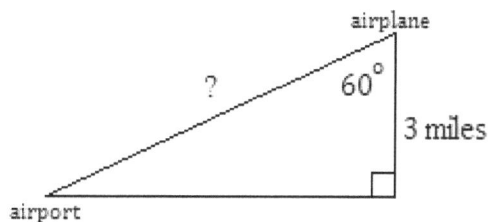

123. In the figure below, the center of the circle is O and $\overline{PQ}$ is tangent to the circle at Q. What is the area of the shaded region to the nearest tenth?

    A. 2.7
    B. 2.9
    C. 3.1
    D. 3.3
    E. 3.5

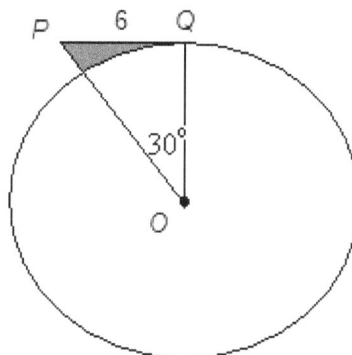

124. A container in the shape of a right circular cylinder has an inside base radius of 5 centimeters and an inside height of 6 centimeters. This cylinder is completely filled with fluid. All of the fluid is then poured into a second right circular cylinder with a larger inside base radius of 7 centimeters. What must be the minimum inside height, in centimeters, of the second container?

    F. $\frac{5}{\sqrt{7}}$

    G. $\frac{7}{5}$

    H. 5

    J. $\frac{150}{49}$

    K. $\frac{25}{6}$

125. Rectangular bricks measuring $\frac{1}{2}$ meter by $\frac{1}{3}$ meter are sold in boxes containing 8 bricks each. What is the least number of boxes of bricks needed to cover a rectangular area that has dimensions 9 meters by 11 meters?

    A.   3
    B.   17
    C.   74
    D.   75
    E.   132

126. Line k contains the point $(4, 0)$ and has slope 5. Which of the following points is on line k ?

    F. $(1, 5)$
    G. $(3, 5)$
    H. $(5, 5)$
    J. $(7, 5)$
    K. $(9, 5)$

127. A line through the origin and $(3, 5)$ is shown in the standard (x, y) coordinate plane below. The acute angle between the line and the positive x-axis has measure θ. What is the value of $\cos \theta$?

    A. $\frac{3}{5}$
    B. $\frac{5}{3}$
    C. $\frac{3}{\sqrt{34}}$
    D. $\frac{\sqrt{34}}{3}$
    E. $\frac{5}{\sqrt{34}}$

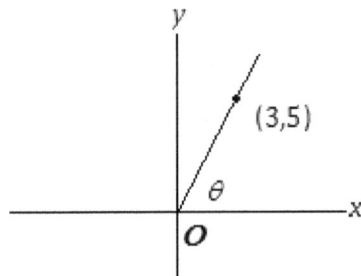

128. The angle of elevation from the tip of the shadow of a 30 meter tall building to the top of the building has a cosine of $\frac{3}{7}$. What is the length of the shadow to the nearest meter?

    F.   12
    G.   13
    H.   14
    J.   15
    K.   16

LEVEL 5: NUMBER THEORY

129. In an empty square field, m rows of m trees are planted so that the whole field is filled with trees. If k of these trees lie along the boundary of the field, which of the following is a possible value for k ?

    A. 14
    B. 49
    C. 86
    D. 125
    E. 276

130. The sum of c and $2d$ is equal to k, and the product of c and $2d$ is equal to j. If j and k are positive numbers, what is $\frac{10}{c} + \frac{5}{d}$ in terms of j and k ?

    F. $\frac{j}{10k}$
    G. $\frac{10k}{j}$
    H. $10jk$
    J. $\frac{10}{j} + \frac{5}{k}$
    K. $\frac{10}{2j+k}$

131. If m and n are positive integers such that the greatest common factor of $m^3 n$ and $m^2 n^2$ is 175, then which of the following could m equal?

    A. 5
    B. 7
    C. 25
    D. 35
    E. 175

132. In the repeating decimal

$$0.\overline{7654321} = 0.765432176543217654321\ldots$$

where the digits 7654321 repeat, which digit is in the 2012th place to the right of the decimal?

    F. 3
    G. 4
    H. 5
    J. 6
    K. 7

137

133. What is the sum of the first 5 terms of the arithmetic sequence in which the 7th term is 10 and the 11th term is 16 ?

    A. 18.5
    B. 20
    C. 22.5
    D. 25
    E. 27.5

134. The first and second terms of a geometric sequence are k and bk, in that order. What is the 500th term of the sequence?

    F. $(bk)^{500}$
    G. $(bk)^{499}$
    H. $b^{501}k$
    J. $b^{500}k$
    K. $b^{499}k$

135. The quantity $\sqrt[n]{5^x}$ is defined when n is an integer greater than 2 and x is any nonzero real number. Which of the following is a relationship between n and x that will always make $\sqrt[n]{5^x}$ a positive integer?

    A. x is less than n

    B. n is less than x

    C. $n + x = 1$

    D. $\frac{n}{x}$ is a positive integer

    E. $\frac{x}{n}$ is a positive integer

136. In the equation $\log_3 63 - \log_3 7 = \log_6 x$, what is the real value of x ?

    F.   7
    G.   9
    H.   36
    J.   63
    K.   199

LEVEL 5: ALGEBRA

137. Given $f(x) = \frac{x-3}{x^2}$, which of the following expressions is equal to $f(x+5)$ for all x in its domain?

    A. $\frac{x+2}{x^2+10x+25}$

    B. $\frac{x+2}{x^2+5}$

    C. $\frac{x+2}{2x+10}$

    D. $\frac{x-3}{x+5}$

    E. $\frac{-x^2+x-3}{x^2}$

138. Consider the function $f(x) = x^2 - 5x + c$. The graph of f in the (x, y) coordinate plane is a parabola whose vertex lies on the x-axis. What is the value of c?

    F. -6.25

    G. -6

    H. 0

    J. 6

    K. 6.25

139. If $3x = 5 + 2y$ and $4x = 3 - 5y$, what is the value of x?

    A. $\frac{23}{31}$

    B. $\frac{37}{41}$

    C. $\frac{31}{23}$

    D. $\frac{41}{25}$

    E. 31

140. The solution set of which of the following equations is the set of real numbers that are 7 units from -2?

    F. $|x + 2| = 7$

    G. $|x - 2| = 7$

    H. $|x + 2| = -7$

    J. $|x - 7| = 2$

    K. $|x + 7| = 2$

141. When $x \neq 7$, $\dfrac{3x}{x^2-49} + \dfrac{3x}{7-x}$ is equivalent to:

    A. $\dfrac{-3x^2-21x}{x^2-49}$

    B. $\dfrac{-21x}{x^2-49}$

    C. $\dfrac{6-21x}{x^2-49}$

    D. $\dfrac{-3x^2-18x}{x^2-49}$

    E. $\dfrac{-3x^2}{x^2-49}$

142. Which of the following matrices is equal to the matrix product $\begin{bmatrix} -3 & 1 \\ 4 & -2 \end{bmatrix} \cdot \begin{bmatrix} 1 \\ -2 \end{bmatrix}$?

    F. $\begin{bmatrix} -3 & -2 \\ 4 & 4 \end{bmatrix}$

    G. $\begin{bmatrix} -2 & 2 \\ 2 & -4 \end{bmatrix}$

    H. $\begin{bmatrix} -3 & 4 \\ -2 & 4 \end{bmatrix}$

    J. $\begin{bmatrix} -1 \\ 0 \end{bmatrix}$

    K. $\begin{bmatrix} -5 \\ 8 \end{bmatrix}$

143. A parabola with axis of symmetry $x = 2$ crosses the x-axis at $(2 + \sqrt{3}, 0)$. At what other point, if any, does the parabola cross the x-axis?

    A. $\left(-2 - \sqrt{3}, 0\right)$
    B. $\left(-2 + \sqrt{3}, 0\right)$
    C. $\left(2 - \sqrt{3}, 0\right)$
    D. No other point
    E. Cannot be determined from the given information

144. The equation $3^{x^2-2x+8} = 243$ has two solutions. Let a be the sum of these solutions and let b be the product of these solutions. What is $a - b$?

    F. -2
    G. -1
    H. 0
    J. 1
    K. 2

LEVEL 5: PROBLEM SOLVING AND DATA

145. A rectangle was changed by increasing its length by r percent and decreasing its width by 20%. If these changes increased the area of the rectangle by 4%, what is the value of r ?

    A. 10
    B. 20
    C. 30
    D. 35
    E. 40

146. What is $\frac{14}{11}$% of $\frac{10}{7}$?

    F. $\frac{1}{550}$
    G. $\frac{1}{55}$
    H. $\frac{98}{110}$
    J. $\frac{20}{11}$
    K. $\frac{200}{11}$

147. If $t = j + k + m + n + p + q + r$, what is the average (arithmetic mean) of j, k, m, n, p, q, r, and t in terms of t ?

    A. $\frac{t}{2}$
    B. $\frac{t}{3}$
    C. $\frac{t}{4}$
    D. $\frac{t}{5}$
    E. $\frac{t}{6}$

148. Let a, b and c be numbers with $a < b < c$ such that the average of a and b is 2, the average of b and c is 4, and the average of a and c is 3. What is the average of a, b and c ?

    F. 1
    G. 2
    H. 3
    J. 4
    K. 5

149. In Dr. Steve's AP Calculus BC class, students are given a grade between 0 and 100, inclusive on each exam. Jason's average (arithmetic mean) for the first 3 exams was 90. What is the lowest grade Jason can receive on his 4th exam and still be able to have an average of 90 for all 7 exams that will be given?

    A. 52
    B. 55
    C. 56
    D. 58
    E. 60

150. At Brilliance University, the chess team has 16 members and the math team has 13 members. If a total of 7 students belong to only one of the two teams, how many students belong to both teams?

    F. 7
    G. 11
    H. 15
    J. 22
    K. 24

151. Seven cards, each of a different color are shuffled and placed in a row. In how many ways can the cards be placed so that the blue card is placed at an end?

    A. 1440
    B. 720
    C. 28
    D. 21
    E. 14

152. One hundred cards numbered 200 through 299 are placed into a bag. After shaking the bag, 1 card is randomly selected from the bag. Without replacing the first card, a second card is drawn. If the first card drawn is 265, what is the probability that both cards drawn have the same tens digit?

    F. $\frac{1}{8}$
    G. $\frac{1}{9}$
    H. $\frac{1}{10}$
    J. $\frac{1}{11}$
    K. $\frac{1}{99}$

LEVEL 5: GEOMETRY

153. The lengths of the sides of an isosceles triangle are 22, m, and m. If m is an integer, what is the smallest possible perimeter of the triangle?

    A. 30
    B. 31
    C. 32
    D. 34
    E. 46

154. An isosceles right triangle, T_1, has a hypotenuse of length $10\sqrt{2}$ units. The vertices of a second right triangle, T_2, are the midpoints of the sides of T_1. The vertices of a third right triangle, T_3, are the midpoints of the sides of T_2. This process continues indefinitely, with the vertices of T_{k+1} being the midpoints of the sides of T_k for each integer $k > 0$. What is the sum of the areas, in square units, of $T_1, T_2, \ldots$?

    F. $\dfrac{25}{3}$

    G. $\dfrac{50}{3}$

    H. $\dfrac{100}{3}$

    J. $\dfrac{200}{3}$

    K. 200

155. A circle in the standard (x, y) coordinate plane is tangent to the x-axis at 3 and tangent to the y-axis at -3. Which of the following is an equation of the circle?

    A. $x^2 + y^2 = 3$
    B. $x^2 - y^2 = 9$
    C. $(x - 3)^2 + (y + 3)^2 = 3$
    D. $(x - 3)^2 + (y + 3)^2 = 9$
    E. $(x + 3)^2 + (y - 3)^2 = 9$

156. In the figure below, AB is the arc of a circle with center O. If the length of arc AB is 4π, what is the area of region OAB to the nearest tenth?

    F. 89.2
    G. 89.5
    H. 89.8
    J. 90.2
    K. 90.5

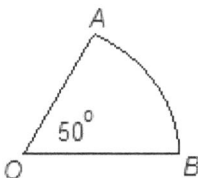

157. For any cube, if the volume is V cubic centimeters and the surface area is S square centimeters, then S is directly proportional to V^n for $n =$

    A. $\frac{1}{2}$

    B. $\frac{2}{3}$

    C. $\frac{3}{2}$

    D. 2

    E. 3

158. In the figure below, $\overline{QS}$ is the shorter diagonal of rhombus $PQRS$ and T is on $\overleftrightarrow{PS}$. The measure of angle PQS is $x°$. What is the measure of RST, in terms of x ?

    F. $x°$

    G. $2x°$

    H. $\frac{1}{2}x°$

    J. $90° - x°$

    K. $180° - (2x)°$

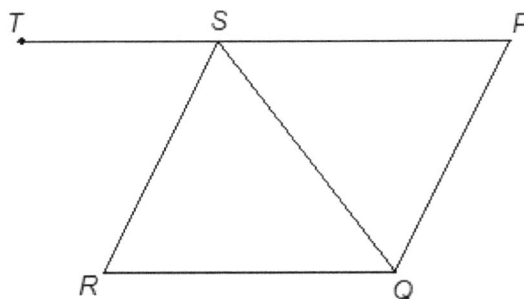

159. For the triangles in the figure below, which of the following ratios of side lengths is equivalent to the ratio of the perimeter of $\triangle CBA$ to the perimeter of $\triangle MAC$?

    A. $AB:CA$
    B. $AB:AM$
    C. $AB:BC$
    D. $AB:CM$
    E. $BC:CM$

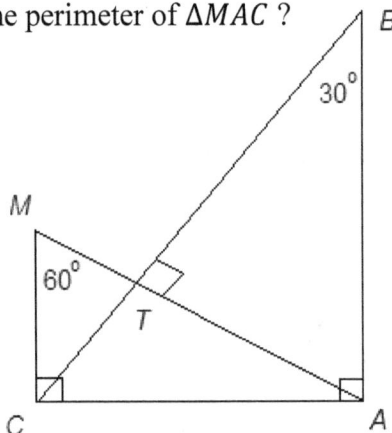

160. What is the degree measure of the largest angle of a triangle that has sides of length 7, 8, and 9 to the nearest degree?

    F. $75°$
    G. $73°$
    H. $17°$
    J. $16°$
    K. $1°$

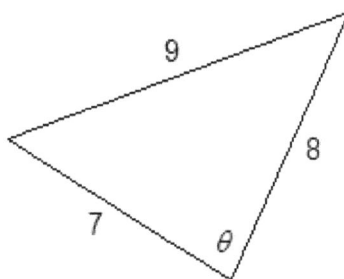

PROBLEMS BY LEVEL AND TOPIC
PROBLEM SET B

Full solutions to these problems are available for free download here:
www.SATPrepGet800.com/UniACTyWB

LEVEL 1: NUMBER THEORY

1. If $a + 7$ is an even integer, then a could be which of the following?

   A. -4
   B. -3
   C. 0
   D. 2
   E. 4

2. On the first Monday in February, Mrs. Green gave her students 7 math problems to solve. On each school day after that, she gave the students 8 math problems to solve. During the first 15 school days, how many math problems had she given the students to solve?

   F. 15
   G. 100
   H. 105
   J. 119
   K. 120

3. Which of the following numbers is less than 0.416 ?

   A. 0.4106
   B. 0.4161
   C. 0.4166
   D. 0.42
   E. 0.421

4. What is the greatest common divisor of 10, 15, 25, and 45 ?

   F. 1
   G. 3
   H. 5
   J. 10
   K. 15

5. The number 0.000 000 000 000 015 is equivalent to which of the following expressions?

    A. 1.5×10^{15}
    B. 1.5×10^{14}
    C. 1.5×10^{13}
    D. 1.5×10^{-14}
    E. 1.5×10^{-15}

6. The second term of an arithmetic sequence is 15 and the third term is 10. What is the first term?

    F. -15
    G. -10
    H. $\frac{1}{15}$
    J. 10
    K. 20

7. For $i = \sqrt{-1}$, the sum $(2 - 3i) + (-5 + 6i)$ is

    A. $-7 + 3i$
    B. $-7 + 9i$
    C. $-3 - 3i$
    D. $-3 + 3i$
    E. $-3 - 18i$

8. $7y^{11} \cdot 4y^{11}$ is equivalent to:

    F. $11y^{22}$
    G. $11y^{121}$
    H. $28y^{11}$
    J. $28y^{22}$
    K. $28y^{121}$

LEVEL 1: ALGEBRA

9. If $4b - 5 = 17$, then $b =$

    A. 4.0
    B. 5.5
    C. 10.0
    D. 17.5
    E. 22.0

10. For which of the following values of x will the value of $11x + 5$ be greater than 27 ?

    F. 3
    G. 3
    H. 1
    J. 0
    K. −1

11. Let the function f be defined as $f(x) = \frac{-x^2+1}{x}$. What is the value of $f(-1)$?

    A. −2
    B. −1
    C. 0
    D. 1
    E. 2

12. If $10xz - 15yz = kz(2x - ny)$ where a and b are positive real numbers, what is the value of kn ?

    F. 8
    G. 12
    H. 15
    J. 18
    K. 30

13. $x^2 - 73x + 27 - 46x^2 + 75x$ is equivalent to:

    A. $-29x^2$
    B. $-29x^6$
    C. $-45x^4 + 2x^2 + 27$
    D. $-45x^2 + 2x + 27$
    E. $-44x^2 + 2x + 27$

14. Which of the following is an equivalent simplified expression for $3(5x + 8) - 4(3x - 2)$?

    F. $x + 16$
    G. $3x + 16$
    H. $3x + 32$
    J. $7x + 2$
    K. $7x + 3$

15. If $a = 4$, $b = 3$, and $c = -7$, then what is the value of $(b + c)(a + b - c)$?

    A. −56
    B. − 4
    C. 0
    D. 4
    E. 56

16. A Celsius temperature C can be approximated by subtracting 32 from the Fahrenheit temperature F and then multiplying by $\frac{1}{2}$. Which of the following expresses this approximation method? (Note: The symbol $\approx$ means "is approximately equal to.")

    F. $C \approx \frac{1}{2}(F - 32)$

    G. $C \approx \frac{1}{2}F - 32$

    H. $C \approx 2(F - 32)$

    J. $C \approx 2F - 32$

    K. $C \approx \sqrt{F} - 32$

LEVEL 1: PROBLEM SOLVING AND DATA

17. At an adoption center, 4 guinea pigs are selected at random from each group of 15. At this rate, how many guinea pigs will be selected in total if the adoption center has 90 guinea pigs?

    A. 4
    B. 6
    C. 12
    D. 24
    E. 48

18. You are about to pay for a hat priced at \$9.99. A sales tax of 9% of \$9.99 will be added (rounded to the nearest cent). You have 15 one-dollar bills, but how much will you need, in cents, if you want to be ready with exact change?

    F. 89
    G. 75
    H. 53
    J. 41
    K. 33

19. The daily totals of shoppers in the Moonlight Convenience Store last week were 97, 123, 216, 26, 41, 187, and 87. What was the average number of shoppers each day?

    A.   60
    B. 111
    C. 155
    D. 259
    E. 775

20. The mean of ten numbers is 70. If the sum of nine of the numbers is 500, what is the tenth number?

    F.   50
    G. 100
    H. 200
    J.   430
    K. 500

21. A menu lists 2 appetizers, 5 meals, 4 drinks, and 3 desserts. A dinner consists of 1 of each of these 4 items. How many different dinners are possible from this menu?

    A.   2
    B.   4
    C. 14
    D. 72
    E. 120

22. A bag contains 12 pieces of fabric. 3 of the pieces are blue, 5 are red, and the remaining are green. Without looking, James reaches into the bag and randomly selects a piece of fabric. What is the probability that the fabric he selects is NOT red?

    F. $\dfrac{3}{12}$
    G. $\dfrac{1}{3}$
    H. $\dfrac{5}{12}$
    J. $\dfrac{7}{12}$
    K. $\dfrac{2}{3}$

23. Of the 30 students in a math class, 7 earned A's, 12 earned B's, 5 earned C's, and 6 earned D's. If a student from the class is chosen at random, what is the probability that the student chosen had earned a B in the class?

    A. $\dfrac{1}{5}$
    B. $\dfrac{2}{5}$
    C. $\dfrac{3}{5}$
    D. $\dfrac{2}{3}$
    E. $\dfrac{4}{5}$

24. A department consisting of 28 faculty members is meeting to choose a chair for the executive committee. The representative, who will be selected at random, CANNOT be any of the 5 faculty members that are already chairs of other committees. What is the probability that Dan, who is NOT the chair of another committee, will be selected?

    F. 0

    G. $\frac{1}{28}$

    H. $\frac{1}{23}$

    J. $\frac{1}{5}$

    K. $\frac{3}{14}$

LEVEL 1: GEOMETRY

25. A point at $(-5,6)$ in the standard (x,y) coordinate plane is shifted up 4 units and right 8 units. What are the coordinates of the new point?

    A. $(-1,14)$
    B. $(-13,10)$
    C. $(-13,2)$
    D. $(3,2)$
    E. $(3,10)$

26. In $\triangle PQR$, the sum of the measures of $\angle P$ and $\angle Q$ is 59°. What is the measure of $\angle R$?

    F. 31°
    G. 59°
    H. 118°
    J. 121°
    K. 131°

27. If the degree measures of an isosceles triangle are 100°, $z°$, and $z°$, what is the value of z ?

    A. 80
    B. 70
    C. 60
    D. 40
    E. 30

28. The radius of circle O is 3 centimeters. What is the area of circle O, in square centimeters?

    F. 3π
    G. 4π
    H. 6π
    J. 9π
    K. 12π

29. In the xy-plane, the point $(5, 0)$ is the center of a circle that has radius 5. Which of the following is NOT a point on the circle?

    A. $(10, \ 0)$
    B. $(10, -5)$
    C. $(\ 5, \ \ 5)$
    D. $(\ 5, -5)$
    E. $(\ 0, \ \ 0)$

30. In rectangle $PQRS$, which of the following must be true about the measures of $\angle PQR$ and $\angle QRS$?

    F. each are 90°
    G. each are less than 90°
    H. each are greater than 90°
    J. they add up to 90°
    K. they add up to 360°

31. In the figure below, adjacent sides meet at right angles and the lengths given are in inches. What is the perimeter of the figure, in inches?

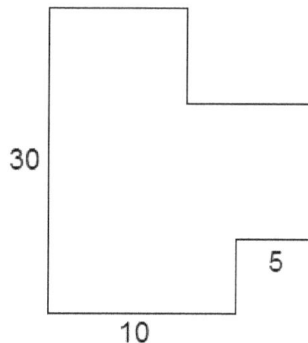

30

5

10

    A. 45
    B. 60
    C. 90
    D. 120
    E. 450

32. In the standard (x, y) coordinate plane, what is the slope of the line segment joining the points $(3, -5)$ and $(7, 2)$?

    F. $-\dfrac{7}{4}$
    G. $-\dfrac{4}{7}$
    H. $\dfrac{4}{7}$
    J. $\dfrac{7}{10}$
    K. $\dfrac{7}{4}$

LEVEL 2: NUMBER THEORY

33. For two consecutive integers, the result of adding the larger integer and twice the smaller integer is 106. What is the larger integer?

    A. 12
    B. 25
    C. 35
    D. 36
    E. 72

34. What is the correct ordering of $\sqrt{2}$, 1.4, and $\frac{\sqrt{5}}{2}$ from least to greatest?

    F. $\sqrt{2} < 1.4 < \frac{\sqrt{5}}{2}$

    G. $1.4 < \sqrt{2} < \frac{\sqrt{5}}{2}$

    H. $\sqrt{2} < \frac{\sqrt{5}}{2} < 1.4$

    J. $1.4 < \frac{\sqrt{5}}{2} < \sqrt{2}$

    K. $\frac{\sqrt{5}}{2} < 1.4 < \sqrt{2}$

35. Which of the following numbers is between $\frac{1}{15}$ and $\frac{2}{25}$?

    A. 0.06
    B. 0.07
    C. 0.09
    D. 0.11
    E. 0.12

36. What is the least common denominator of the fractions $\frac{3}{10}$, $\frac{2}{45}$, and $\frac{5}{27}$?

    F. 54
    G. 270
    H. 450
    J. 2430
    K. 12,150

37. What is the greatest positive integer that is a divisor of 26, 39, 117, and 169 ?

    A. 1
    B. 3
    C. 5
    D. 13
    E. 26

38. For all positive integers m, what is the greatest common factor of the 3 numbers $40m$, $60m$, and $100m$?

    F. 20
    G. 40
    H. m
    J. 20m
    K. 40m

39. When we subtract $-3 + i$ from $5 - 4i$ we get which of the following complex numbers?

    A. $2 - 3i$
    B. $2 - 5i$
    C. $-8 - 5i$
    D. $-8 + 5i$
    E. $8 - 5i$

40. $(y^7)^4$ is equivalent to:

    F. y^{28}
    G. y^{11}
    H. $4y^8$
    J. $4y^7$
    K. $28y$

LEVEL 2: ALGEBRA

41. Which of the following ordered pairs (x, y) does not satisfy the inequality $7x - 2y < 3$?

    A. $(1, \ 4)$
    B. $(2, 10)$
    C. $(3, 11)$
    D. $(4, 12)$
    E. $(5, 18)$

42. The functions f and g are defined below. What is the value of $f(8) - g(2)$?

$$f(x) = 2x - 5$$
$$g(x) = x^2 + 3x - 2$$

    F. 0
    G. 1
    H. 2
    J. 3
    K. 4

43. If $x - y = 7$ and $\frac{x}{5} = 3$, what is the value of $x + y$?

    A. 18
    B. 20
    C. 21
    D. 22
    E. 23

44. What polynomial must be added to $x^2 + 3x - 5$ so that the sum is $5x^2 - 8$?

    F. $4x^2 - 5x + 6$
    G. $4x^2 - 3x - 3$
    H. $5x^2 - 3x - 3$
    J. $5x^2 - 3x + 6$
    K. $6x^2 + 3x - 13$

45. Which of the following expressions is equal to the expression below for all real values of x ?

$$(5x^2 - 3x - 2) - (-x^2 + 2x + 5)$$

    A. $4x^2 - 5x - 7$
    B. $4x^2 - 5x + 3$
    C. $6x^2 - 5x - 7$
    D. $6x^2 - 5x + 3$
    E. $6x^2 - 5x - 3$

46. Which of the following expressions is equivalent to 11 less than the product of x and y?

    F. $x + y = 11$
    G. $xy - 11$
    H. $11xy$
    J. $11(x + y)$
    K. $(x + 11)y$

47. A rectangle has area A, length x and width y. Which of the following represents y in terms of A and x ?

    A. $y = \frac{A}{2x}$
    B. $y = \frac{A}{x}$
    C. $y = \frac{2A}{x}$
    D. $y = \frac{\sqrt{A}}{x}$
    E. $y = \frac{\sqrt{A}}{2x}$

48. If $(x - 2)^2 = 16$, and $x < 0$, what is the value of x ?

    F. -33
    G. -9
    H. -3
    J. -2
    K. -1

LEVEL 2: PROBLEM SOLVING AND DATA

49. On planet Puro, if each month has 12 days and each day has 8 hours, how many full Puro months will have passed after 400 hours?

    A. One
    B. Two
    C. Three
    D. Four
    E. Five

50. A census was given to determine information about the number of children from households in a small community. 100 families were surveyed and the results are shown in the table below.

Number of Children Per Household

Number of Children	0	1	2	3	More than 3
Households with that number of children	11	21	35	17	16

What percent of the households that were surveyed have at least 3 children?

    F. 8%
    G. 16%
    H. 17%
    J. 33%
    K. 37%

51. What is the average (arithmetic mean) of $9 - k$, 9, and $9 + k$?

    A. 3
    B. 9
    C. 15
    D. $3 + \dfrac{k}{3}$
    E. $9 + \dfrac{k}{3}$

155

52. A data set contains 6 elements and has a mean of 5. Five of the elements are 2, 4, 6, 8, and 10. Which of the following is the sixth element?

    F. 0
    G. 1
    H. 2
    J. 3
    K. 4

53. Greg earned scores of 63, 72, 86, and 91 on his first 4 math tests. What is the minimum score Greg needs to earn on his 5th test so that the mean of his scores on all 5 tests is at least 3 points more than the mean of his scores on the first 4 tests?

    A. 90
    B. 91
    C. 92
    D. 93
    E. 94

54. The average of 7 distinct scores has the same value as the median of the 7 scores. The sum of the 7 scores is 126. What is the sum of the 6 scores that is NOT the median?

    F.   18
    G.   52
    H.   96
    J. 108
    K. 118

55. Of the marbles in a jar, 14 are green. Joseph randomly takes one marble out of the jar. If the probability is $\frac{7}{8}$ that the marble he chooses is green, how many marbles are in the jar?

    A. 8
    B. 10
    C. 12
    D. 14
    E. 16

56. The Math Club needs to select 4 officers by first selecting the president, then the vice president, then the secretary, and finally the treasurer. If there are 20 members who are eligible to hold office and no member can hold more than 1 office, which of the following gives the number of different possible results of the election?

    F. $20 \cdot 19 \cdot 18 \cdot 17$
    G. $19 \cdot 18 \cdot 17 \cdot 16$
    H. 20^4
    J. 19^4
    K. 15^4

LEVEL 2: GEOMETRY

57. C is the midpoint of line segment AB, and D and E are the midpoints of AC and CB, respectively. If the length of DE is 7, what is the length of AB ?

    A. 3.5
    B. 7
    C. 10.5
    D. 14
    E. 17.5

58. Four points, $X, Y, Z,$ and W, lie on a circle having a circumference of 30 units. Y is 4 units counterclockwise from X. Z is 10 units clockwise from X. W is 14 units clockwise from X and 16 units counterclockwise from X. What is the order of the points, starting with X and going clockwise around the circle?

    F. X, Y, Z, W
    G. X, Y, W, Z
    H. X, Z, Y, W
    J. X, Z, W, Y
    K. X, Z, Y, W

59. Given right triangle $\triangle PQR$ below, what is the length of $\overline{PQ}$?

    A. $\sqrt{2}$
    B. $\sqrt{5}$
    C. 5
    D. 7
    E. 11

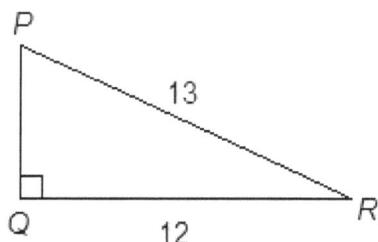

60. In $\triangle ABD$ below, if $z = 37$, what is the value of y ?

    F. 38
    G. 75
    H. 90
    J. 100
    K. 105

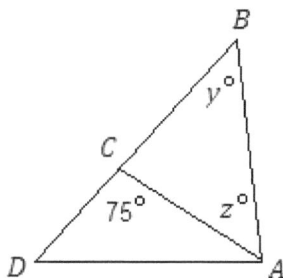

157

61. For all triangles $\triangle ABC$ where the measure of angle B is greater than the measure of angle A, such as the triangle shown below, which of the following statements is true?

    A. Sometimes $AC < BC$ and sometimes $AC = BC$
    B. Sometimes $AC > BC$ and sometimes $AC = BC$
    C. $AC < BC$
    D. $AC = BC$
    E. $AC > BC$

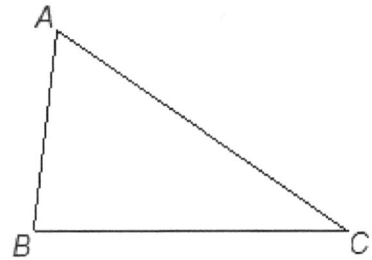

62. What is the slope of any line perpendicular to the line $3x - 2y = 5$?

    F. -2

    G. $-\dfrac{3}{5}$

    H. $-\dfrac{2}{3}$

    J. 3

    K. 5

63. If $0 \leq x \leq 90°$ and $\cos x = \dfrac{5}{13}$, then $\tan x =$

    A. $\dfrac{5}{12}$

    B. $\dfrac{12}{13}$

    C. $\dfrac{13}{12}$

    D. $\dfrac{12}{5}$

    E. $\dfrac{13}{5}$

64. In the figure below, lines j and k are parallel and lines ℓ and m are parallel. If the measure of $\angle 1$ is 68°, what is the measure of $\angle 2$?

    F. 158°
    G. 112°
    H. 98°
    J. 76°
    K. 68°

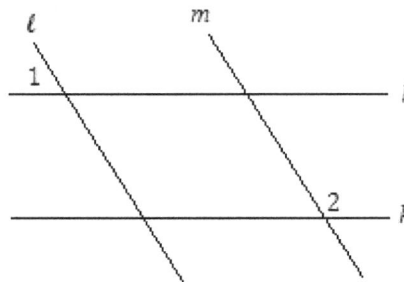

158

LEVEL 3: NUMBER THEORY

65. If k and m are even integers, which of the following expressions must be an odd integer?

    A. km
    B. $5km$
    C. $km + 1$
    D. $\dfrac{k+m}{2}$
    E. $k^2 + m$

66. John purchased a house for \$340,000. He financed all of the \$340,000 and started loan payments of \$2100 a month for 30 years. At the end of the 30-year period, how much more than the purchase price will John have paid for his house?

    F. \$525,000
    G. \$416,000
    H. \$340,000
    J. \$160,000
    K. \$102,250

67. The number line graph below is the graph of which of the following inequalities?

    A. $-3 \le x$ and $5 \le x$
    B. $-3 \le x$ and $5 \ge x$
    C. $-3 \le x$ or $5 \le x$
    D. $-3 \ge x$ or $5 \le x$
    E. $-3 \ge x$ or $5 \ge x$

68. Which of the following is a rational number?

    F. $\sqrt{3}$
    G. $\sqrt{\pi^2}$
    H. $\sqrt{11}$
    J. $\sqrt{\dfrac{9}{64}}$
    K. $\sqrt{\dfrac{49}{5}}$

69. What is the least common multiple of the numbers 2, 3, 4, 5, 6, and 7 ?

    A.   1
    B. 105
    C. 210
    D. 420
    E. 5040

70. If $7^x = 6$, then $7^{3x} =$

    F.   2
    G.   9
    H. 18
    J. 36
    K. 216

71. Which of the following expressions is equivalent to $\frac{(2y)^3}{y^7}$?

    A. $\frac{2}{y^4}$
    B. $\frac{6}{y^4}$
    C. $\frac{8}{y^4}$
    D. $2y^{10}$
    E. $6y^{10}$

72. In the equation $\log_2 4 + \log_2 2 = 3 \log_2 x$, the value of x is

    F. 1
    G. 2
    H. 3
    J. 4
    K. 5

LEVEL 3: ALGEBRA

73. The value of x that will make $\frac{x}{3} - 2 = -\frac{11}{4}$ a true statement lies between which of the following numbers?

    A. -3 and -2
    B. -2 and -1
    C. -1 and 0
    D. 0 and 1
    E. 1 and 2

74. Suppose that $h(x) = 4x - 5$ and $h(b) = 17$. What is the value of b ?

    F. 4
    G. 5.5
    H. 10
    J. 15
    K. 17.5

75. A small hotel has 15 rooms which are all occupied. If each room is occupied by either one or two guests and there are 27 guests in total, how many rooms are occupied by two guests?

    A. 6
    B. 10
    C. 12
    D. 15
    E. 27

76. Which of the following expressions is a factor of the polynomial $x^2 - x - 90$?

    F. $x - 7$
    G. $x - 8$
    H. $x - 9$
    J. $x - 10$
    K. $x - 11$

77. Which of the following is equivalent to the following expression?
$$b^2 - 4a^2$$

    A. $(b - 2a)^2$
    B. $(b + 2a)^2$
    C. $(b - a)(b + 4a)$
    D. $(b - 2a)(b + 2a)$
    E. $b - 2a$

78. What is the determinant of the matrix $\begin{bmatrix} 2 & -8 \\ 1 & -5 \end{bmatrix}$?

    F. -18
    G. -2
    H. 4
    J. 2
    K. 16

79. If $3x^2 + 9x = 84$, what are the possible values for x ?

    A. -4 and 7
    B. -7 and 4
    C. -7 and -4
    D. -7 and -12
    E. 12 and 14

80. What is the sum and product of the two solutions of the equation $x^2 - 3x + 8 = 0$?

    F. sum = -3,   product =   8
    G. sum =   3,   product =   8
    H. sum =   3,   product = -8
    J. sum = -8,   product = -3
    K. sum =   8,   product =   3

LEVEL 3: PROBLEM SOLVING AND DATA

81. John drove 8 hours from New York to Virginia. The total distance he travelled was 535 miles, and he averaged 65 miles per hour for the first 3 hours. Which of the following is closest to his average driving speed, in miles per hour, for the remainder of his drive?

    A. 60
    B. 62
    C. 65
    D. 66
    E. 68

82. What number k must be added to both the numerator and the denominator of the fraction $\frac{5}{19}$ to make the resulting fraction equal to $\frac{6}{13}$?

    F. 8
    G. 7
    H. 6
    J. 5
    K. 4

83. Karen, Lisa, and Maria shared a pizza pie. Karen ate $\frac{1}{4}$ of the pie, Lisa ate $\frac{1}{3}$ of the pie, and Maria ate the rest. What is the ratio of Karen's share to Lisa's share to Maria's share?

    A. $3:4:5$
    B. $3:4:6$
    C. $4:3:5$
    D. $4:3:6$
    E. $5:4:3$

84. Daniel and eight other students took two exams, and each exam yielded an integer grade for each student. The two grades for each student were added together. The sum of these two grades for each of the nine students was 150, 183, 100, 126, 151, 171, 106, 164, and Daniel's sum, which was the median of the nine sums. If Daniel's first test grade was 70, what is one possible grade Daniel could have received on the second test?

    F. 76
    G. 78
    H. 80
    J. 82
    K. 84

85. Between Town A and Town B there are 5 roads, between Town B and Town C there are 2 roads, and between Town C and Town D there are 4 roads. If a traveler were to travel from Town A to Town D, passing first through B, then through C, how many different routes does he have to choose from?

    A. 11
    B. 20
    C. 40
    D. 60
    E. 80

86. Shown below, a circular board with a spinner has 3 regions (white, black, and grey) whose areas are in the ratio of $1:3:4$, repectively. The spinner is spun and it lands in one of the three regions at random. What is the probability that the region it lands in is NOT the white region?

    F. $\frac{1}{8}$

    G. $\frac{3}{8}$

    H. $\frac{1}{2}$

    J. $\frac{5}{8}$

    K. $\frac{7}{8}$

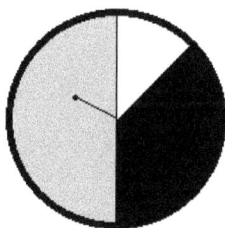

87. A chemist is testing 7 different liquids. For each test, the chemist chooses 4 of the liquids and mixes them together. What is the least number of tests that must be done so that every possible combination of liquids is tested?

    A. 3
    B. 8
    C. 11
    D. 28
    E. 35

88. Set S contains only the integers 0 through 200 inclusive. If a number is selected at random from S, what is the probability that the number selected will be greater than 175?

    F. $\frac{8}{67}$

    G. $\frac{25}{201}$

    H. $\frac{1}{8}$

    J. $\frac{5}{38}$

    K. $\frac{3}{25}$

LEVEL 3: GEOMETRY

89. In the figure below, $k \parallel n$. What is the value of x?

A. 30
B. 40
C. 50
D. 60
E. 70

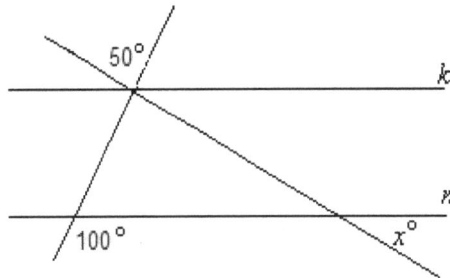

90. In $\triangle ABC$, the length of $\overline{AB}$ is $\sqrt{82}$ centimeters, and the length of $\overline{BC}$ is $\sqrt{71}$ centimeters. If it can be determined, what is the length, in centimeters, of $\overline{AC}$?

F. $\sqrt{11}$
G. $\sqrt{71}$
H. $\sqrt{82}$
J. $\sqrt{153}$
K. Cannot be determined from the given information

91. A circle in the standard (x, y) coordinate plane has equation $(x - 5)^2 + (y + 2)^2 = 73$. What are the radius of the circle, in coordinate units, and the coordinates of the center of the circle?

A. $\sqrt{73}$, $(-5, \ 2)$
B. $\sqrt{73}$, $(\ 5, -2)$
C. $\sqrt{73}$, $(-2, \ 5)$
D. 73, $(\ 2, -5)$
E. 73, $(-5, \ 2)$

92. A small capsule is created from two congruent right circular cones and a right circular cylinder with measurements shown in the figure below. Of the following, which is closest to the volume of the capsule, in cubic inches?

F. 0.39
G. 0.98
H. 1.18
J. 2.16
K. 1.37

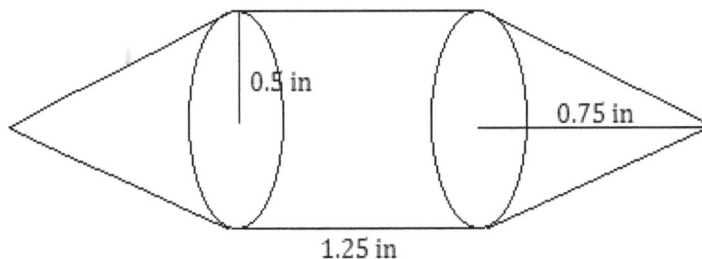

93. A rectangle has a perimeter of 16 meters and an area of 15 square meters. What is the longest of the side lengths, in meters, of the rectangle?

    A. 3
    B. 5
    C. 10
    D. 15
    E. 16

94. Which of the following is an equation of the line in the xy-plane that passes through the point $(0, -3)$ and is parallel to the line $y = -4x + 7$?

    F. $4x + y = -6$
    G. $4x + y = -3$
    H. $4x + y = 3$
    J. $-x + 4y = 6$
    K. $-x + 4y = 3$

95. As shown below, a 10-foot ramp forms an angle of 23° with the ground, which is horizontal. Which of the following is an expression for the vertical rise, in feet, of the ramp?

    A. $10 \cos 23°$

    B. $10 \sin 23°$

    C. $10 \tan 23°$

    D. $10 \cot 23°$

    E. $10 \sec 23°$

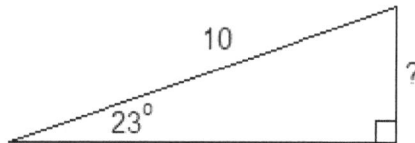

96. Given that $\cos^2 x = \frac{5}{16}$, what is $\sin^2 x$?

    F. $\frac{11}{16}$

    G. $\frac{16}{11}$

    H. $\frac{5}{11}$

    J. $\frac{11}{5}$

    K. $\frac{16}{5}$

LEVEL 4: NUMBER THEORY

97. The sum of 15 positive odd integers is 67. Some of these integers are equal to each other. What is the greatest possible value of one of these integers?

    A. 37
    B. 49
    C. 51
    D. 53
    E. 55

165

98. Jason cut a piece of paper into 5 equal pieces. He threw one piece away, and cut each of the remaining pieces into 4 equal pieces. He again threw one of these pieces away, and cut each of the remaining pieces into 3 equal pieces. He again threw one piece away and cut each of the remaining pieces into 2 equal pieces. After throwing 1 more piece away, how many pieces of paper does Jason have left in total?

    F. 43
    G. 44
    H. 86
    J. 87
    K. 88

99. Which of the following numbers is between $\frac{5}{16}$ and $\frac{2}{5}$?

    A. $\frac{31}{100}$
    B. $\frac{39}{125}$
    C. $\frac{57}{160}$
    D. $\frac{101}{250}$
    E. $\frac{41}{100}$

100. When the positive integer k is divided by 14 the remainder is 4. When the positive integer m is divided by 14 the remainder is 9. What is the remainder when the product km is divided by 7 ?

    F. 0
    G. 1
    H. 2
    J. 3
    K. 4

101. A tennis ball is dropped from 567 centimeters above the ground and after the fourth bounce it rises to a height of 7 centimeters. If the height to which the tennis ball rises after each bounce is always the same fraction of the height reached on its previous bounce, what is this fraction?

    A. $\frac{1}{81}$
    B. $\frac{1}{27}$
    C. $\frac{1}{9}$
    D. $\frac{1}{3}$
    E. $\frac{1}{2}$

102. In the complex numbers, where $i^2 = -1$, $(3 - 5i)^2 =$

    F. -16
    G. 34
    H. $9 - 25i$
    J. $9 + 25i$
    K. $-16 - 30i$

103. If k and r are positive rational numbers such that $k^{5r} = 4$, then $k^{15r} = $?

    A.   8
    B.   12
    C.   16
    D.   60
    E.   64

104. The value of $\log_5 \left(25^{\frac{7}{4}} \right)$ is between which of the following pairs of consecutive integers?

    F.   0 and 1
    G.   2 and 3
    H.   3 and 4
    J.   5 and 6
    K.   6 and 7

LEVEL 4: ALGEBRA

105. The equation shown below is true for what value of a ?
$$5(a - 3) - 3(a - 2) = 7a$$

    A. $-\dfrac{5}{9}$

    B. $-\dfrac{9}{5}$

    C. 0

    D. $\dfrac{9}{5}$

    E. $\dfrac{5}{9}$

106. The figure below shows the graph of the function h and line segment $\overline{AB}$, which has a y-intercept of $(0, b)$. For how many values of x between j and k does $h(x) = b$?

F. Zero
G. One
H. Two
J. Three
K. Four

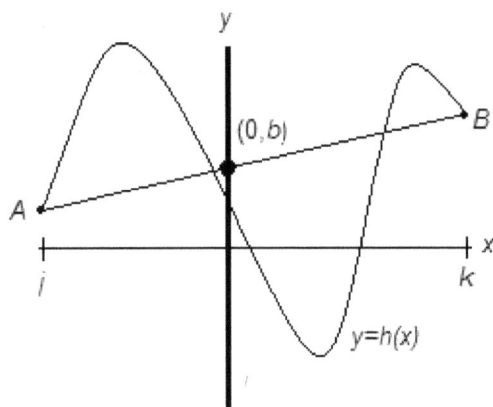

107. In the xy-plane, the graph of the function h, with equation $h(x) = ax^2 - 16$, passes through the point $(-2, 4)$. What is the value of a ?

A. -2
B. 4
C. 5
D. 20
E. 24

108. The graph of $y = \frac{1-x}{2+x}$ in the standard (x, y) coordinate planc has an asymptote with which of the following equations?

F. $x = -1$
G. $x = 1$
H. $x = 2$
J. $y = -1$
K. $y = \frac{1}{2}$

109. Which of the following expressions is a factor of $x^3 - 125$?

A. $x - 5$
B. $x + 5$
C. $x + 125$
D. $x^2 + 25$
E. $x^2 - 5x + 25$

110. A group of friends will rent out a hotel for \$1800 for a party. The cost of the hotel will be equally distributed among the friends who plan to attend the party. The current cost per person will increase by \$15 if 6 of the friends decide not to attend the party. How many friends are currently planning to attend the party?

F. 10
G. 20
H. 24
J. 30
K. 42

111. Given that $\begin{bmatrix} a & b \\ c & 7 \end{bmatrix} = k \begin{bmatrix} 1 & 2 \\ 3 & 4 \end{bmatrix}$ for some real number k, what is $2ab$?

    A. $\frac{7}{4}$

    B. $\frac{7}{2}$

    C. $\frac{49}{4}$

    D. $\frac{49}{2}$

    E.   49

$$y = -5(x - 3)^2 + 2$$

112. In the xy-plane, line ℓ passes through the point $(-1,5)$ and the vertex of the parabola with the equation above. What is the slope of line ℓ ?

    F. $-\frac{4}{3}$

    G. $-\frac{3}{4}$

    H.   0

    J. $\frac{3}{4}$

    K. $\frac{4}{3}$

LEVEL 4: PROBLEM SOLVING AND DATA

113. If the ratio of two positive integers is 7 to 6, which of the following statements about these integers CANNOT be true?

    A.   Their sum is an even integer.
    B.   Their sum is an odd integer.
    C.   Their product is divisible by 11.
    D.   Their product is an even integer.
    E.   Their product is an odd integer.

114. If $z > 0$, then 3 percent of 8 percent of $2z$ equals what percent of z ?

    F.   0.048
    G.   0.48
    H.   4.8
    J.   13.6
    K.   48

115. The average (arithmetic mean) age of the people in a certain group was 32 years before one of the members left the group and was replaced by someone who is 10 years younger than the person who left. If the average age of the group is now 30 years, how many people are in the group?

    A. 1
    B. 2
    C. 3
    D. 4
    E. 5

116. The graph below shows the frequency distribution of a list of randomly generated integers between 0 and 6. Which of the following statements about the mean of the list of integers is true?

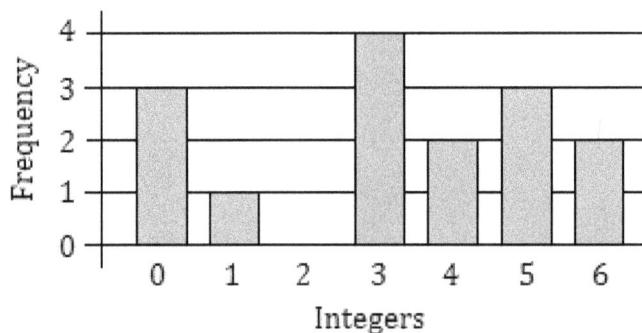

    F. The mean is less than 3.
    G. The mean is 3.
    H. The mean is between 3 and 4.
    J. The mean is 4.
    K. The mean is greater than 4.

117. In a game of catch, 5 people stand in a circle. One person (the thrower) throws the ball to another person (the catcher). The thrower cannot throw the ball to the person immediately to their left or right, or to the person who last threw the ball. When will the first person who throws the ball become the catcher for the first time?

    A. 3rd
    B. 4th
    C. 5th
    D. 6th
    E. 10th

118. Which of the following statements is logically equivalent to the following statement?

             If a spider is an insect, then it has eight legs.

    F. If a spider is not an insect, then it has less than eight legs.
    G. If a spider does not have eight legs, then it is not an insect.
    H. A spider has eight legs if and only if it is an insect.
    J. If a spider does not have eight legs, then it is not an insect.
    K. If a spider has more than eight legs, then it is an insect.

170

119. Any 2 points determine a line. If there are 18 points in a plane, no 3 of which lie on the same line, how many lines are determined by pairs of these 18 points?

    A.   35
    B.   36
    C. 153
    D. 306
    E. 4896

120. The x- and y-coordinates of point A are each to be chosen at random from the set of integers -2 through 10. What is the probability that A will be in quadrant III?

    F. $\dfrac{4}{13}$
    G. $\dfrac{5}{13}$
    H. $\dfrac{4}{169}$
    J. $\dfrac{7}{169}$
    K. $\dfrac{20}{169}$

LEVEL 4: GEOMETRY

121. In the triangle below, $RS = RT = 10$ and $ST = 12$. What is the area of the triangle?

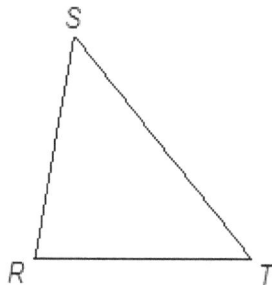

    A. 24
    B. 30
    C. 48
    D. 60
    E. 80

122. In the (x, y) coordinate plane, which of the following is an equation of the circle having the points $(3, -3)$ and $(-5, 5)$ as endpoints of a diameter?

    F. $(x + 1)^2 + (y - 1)^2 = \sqrt{32}$
    G. $(x + 1)^2 + (y + 1)^2 = \sqrt{32}$
    H. $(x - 1)^2 + (y + 1)^2 = \sqrt{32}$
    J. $(x - 1)^2 + (y + 1)^2 = 32$
    K. $(x + 1)^2 + (y - 1)^2 = 32$

123. Cube X has surface area A. The edges of cube Y are 4 times as long as the edges of cube X. What is the surface area of cube Y in terms of A ?

    A. $2A$
    B. $4A$
    C. $8A$
    D. $16A$
    E. $64A$

124. Mike has identical containers each in the shape of a cone with internal diameter of 5 inches. He pours liquid from a half-gallon bottle into each container until it is full. If the height of liquid in each container is 8 inches, what is the largest number of full containers that he can pour a half-gallon of liquid? (Note: There are 231 cubic inches in 1 gallon.)

    F. 5
    G. 4
    H. 3
    J. 2
    K. 1

125. Let P be the perimeter of the figure below in meters, and let A be the area of the figure below in square meters. What is the value of $P + A$?

    A. 25
    B. 30
    C. 35
    D. 40
    E. 45

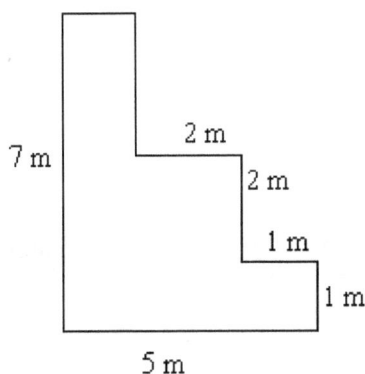

126. What are all real values of a, if any, such that any line through the points $(5, 2)$ and $(a, 2)$ will be horizontal when graphed in the standard (x, y) coordinate plane?

    F. -5
    G. 2
    H. All real numbers except 5 satisfy this condition.
    J. All real numbers satisfy this condition.
    K. No real numbers satisfy this condition.

127. A 10-foot ladder is leaning against a wall such that the angle relative to the level ground is 50°. Which of the following expressions involving cosine gives the distance, in feet, from the base of the ladder to the wall?

    A. $\dfrac{10}{\cos 50°}$
    B. $\dfrac{\cos 50°}{10}$
    C. $\dfrac{1}{10 \cos 50°}$
    D. $10 \cos 50°$
    E. $\cos(10 \cdot 50°)$

128. In the triangle shown below, $\sin x =$

    F. $\dfrac{7}{8 \sin 40°}$

    G. $\dfrac{8}{7 \sin 40°}$

    H. $\dfrac{7 \sin 40°}{8}$

    J. $\dfrac{8 \sin 40°}{7}$

    K. $(8)(7) \sin 40°$

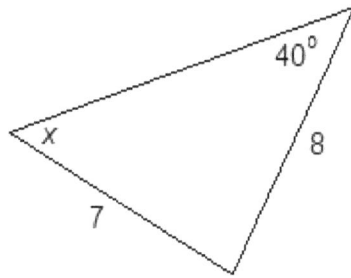

LEVEL 5: NUMBER THEORY

129. In how many of the integers from 1 to 150 does the digit 7 appear at least once?

    A. 14
    B. 15
    C. 23
    D. 24
    E. 25

130. For positive real numbers x, y, and z such that $3x = \dfrac{y\sqrt{2}}{3} = \dfrac{z\sqrt{2}}{3.1}$, which of the following is true?

    F. $x < y < z$
    G. $x < z < y$
    H. $y < x < z$
    J. $y < z < x$
    K. $z < y < x$

131. For all real numbers x, y, and z, with $z \neq 0$, such that the product of x and y is z, which of the following expressions represents the sum of y and z in terms of x and z ?

    A. $x + z$

    B. $xz + z$

    C. $z(x + z)$

    D. $\dfrac{xz+z}{x}$

    E. $\dfrac{x}{z} + z$

132. The integer k is equal to m^2 for some integer m. If k is divisible by 20 and 24, what is the smallest possible positive value of k ?

    F. 44
    G. 240
    H. 480
    J. 2500
    K. 3600

173

133. What is the sum of the first 50 terms of the arithmetic sequence in which the 3rd term is 10 and the 15th term is 22 ?

    A. 125
    B. 275
    C. 525
    D. 875
    E. 1625

134. The first two numbers of a sequence are 5 and 7, respectively. The third number is 12, and, in general, every number after the second is the sum of the two numbers immediately preceding it. How many of the first 400 numbers in the sequence are odd?

    F. 268
    G. 267
    H. 266
    J. 134
    K. 133

135. The sum of an infinite geometric series with first term a and common ratio r with $-1 < r < 1$ is given by $\frac{a}{1-r}$. The sum of a given infinite geometric series is 160 and the common ratio is $\frac{1}{4}$. What is the third term of this series?

    A. 7.5
    B. 12
    C. 17.5
    D. 22
    E. 27.5

136. For the complex number i and an integer k, which of the following is a possible value of i^{2k+1} ?

    F. -1
    G. 0
    H. 1
    J. $1-i$
    K. $-i$

LEVEL 5: ALGEBRA

137. If $9 \le x \le 15$ and $3 \le y \le 5$, what is the greatest possible value of $\frac{7}{x-y}$?

    A. $\frac{7}{12}$
    B. $\frac{7}{10}$
    C. $\frac{7}{9}$
    D. $\frac{7}{6}$
    E. $\frac{7}{4}$

138. Consider the functions $f(x) = \sqrt{x-1}$ and $g(x) = ax + b$. In the standard (x,y) coordinate plane, $y = f(g(x))$ passes through $(0, -1)$ and $(2,3)$. What is the value of $a + b$?

    F. 1
    G. 2
    H. 4
    J. 5
    K. 6

139. If $f(x) = g(x) - h(x)$, where $g(x) = 8x^2 + 13x - 17$ and $h(x) = 8x^2 - 9x + 16$, then $f(x)$ is <u>always</u> divisible by which of the following?

    A. 5
    B. 7
    C. 9
    D. 11
    E. 13

140. The domain of $\dfrac{17}{x^3 - 16x}$ is the set of all real numbers EXCEPT:

    F. $-\dfrac{17}{16}$
    G. 4
    H. -4 and 4
    J. 0 and 4
    K. -4, 0, and 4

141. What is the solution set of $|3x - 2| > 4$?

    A. $\{x \mid x < 2\}$
    B. $\{x \mid x > 2\}$
    C. $\{x \mid x < -\frac{2}{3} \text{ or } x > 2\}$
    D. $\{x \mid x < -2 \text{ or } x > 2\}$
    E. the empty set

142. For all x in the domain of the function $\dfrac{x+2}{x^3-x}$, this function is equivalent to:

    F. $\dfrac{1}{x^2-1} + \dfrac{2}{x^3-x}$
    G. $\dfrac{2}{x^2-1}$
    H. $\dfrac{1}{x^2-1}$
    J. $\dfrac{1}{x-1}$
    K. $\dfrac{1}{x+1}$

143. In the equation $x^2 - bx + c = 0$, b and c are integers. The solutions of this equation are 2 and 3. What is $b - c$?

    A. -11
    B. -1
    C. 1
    D. 5
    E. 11

144. Let the function g be defined by $g(x) = a(x - h)^2$, where h is a positive constant, and a is a negative constant. For what value of x will the function g have its maximum value?

    F. $-h$
    G. $-a$
    H. 0
    J. a
    K. h

LEVEL 5: PROBLEM SOLVING AND DATA

Questions 145 – 146 refer to the following information.

At the beginning of July, 58 percent of the animals in a shelter were dogs, and the rest were cats. By the end of July, 45 percent of the dogs and 63 percent of the cats were adopted.

145. What percentage of the animals in the shelter were adopted?

    A. 5.26
    B. 47.44
    C. 52.56
    D. 71.25
    E. 94.74

146. * What percentage of the animals that were adopted were cats? Round your answer to the nearest tenth.

    F. 26.5
    G. 50.3
    H. 52.6
    J. 54.2
    K. 62.4

147. If x is the average (arithmetic mean) of a and b, y is the average of $2a$ and $3b$, and z is the average of $4a$ and $5b$, what is the average of x, y, and z in terms of a and b ?

    A. $\frac{a}{2} + b$

    B. $\frac{a+b}{2}$

    C. $\frac{7a}{3} + 3b$

    D. $\frac{7a+9b}{6}$

    E. $\frac{7a+9b}{12}$

148. A group of students take a test and the average score is 65. One more student takes the test and receives a score of 92 increasing the average score of the group to 68. How many students were in the initial group?

    F. 5
    G. 6
    H. 7
    J. 8
    K. 9

149. A set of marbles contains only black marbles, white marbles, and yellow marbles. If the probability of randomly choosing a black marble is $\frac{1}{14}$ and the probability of randomly choosing a white marble is $\frac{3}{4}$, what is the probability of randomly choosing a yellow marble?

    A. $\frac{5}{28}$

    B. $\frac{3}{14}$

    C. $\frac{1}{4}$

    D. $\frac{2}{7}$

    E. $\frac{9}{28}$

150. A six-digit number is to be formed using each of the digits 1, 2, 3, 4, 5 and 6 exactly once. How many such numbers are there in which the digits 2 and 3 are next to each other?

    F. 30
    G. 60
    H. 120
    J. 240
    K. 720

151. Tammy and Elizabeth are playing a game where three coins are flipped. Tammy will be awarded 5 points for each of the three coins that shows heads. Let the random variable x represent the total number of points that Tammy receives on any flip of the coins. What is the expected value of x ?

    A. 2.5
    B. 5
    C. 7.5
    D. 12
    E. 15

152. The integers 1 through 6 are written on each of six cards. The cards are shuffled and one card is drawn at random. That card is then replaced, the cards are shuffled again and another card is drawn at random. This procedure is repeated one more time (for a total of three times). What is the probability that the sum of the numbers on the three cards drawn was 3, 4 or 5 ?

    F. $\dfrac{9}{216}$
    G. $\dfrac{10}{216}$
    H. $\dfrac{11}{216}$
    J. $\dfrac{12}{216}$
    K. $\dfrac{13}{216}$

LEVEL 5: GEOMETRY

153. If x is an integer less than 6, how many different triangles are there with sides of length 5, 9 and x ?

    A. One
    B. Two
    C. Three
    D. Four
    E. Five

154. A square is inscribed in a circle of diameter d. What is the perpendicular distance from the center of the circle to a side of the square, in terms of d ?

    F. $\dfrac{d}{2}$
    G. $\dfrac{d\sqrt{2}}{4}$
    H. $\dfrac{d\sqrt{2}}{2}$
    J. d
    K. $d\sqrt{2}$

155. The side lengths of ΔPQR, shown in the figure below, are in inches. One of the 5 points, $V, W, X, Y,$ or Z is the center of a circle that goes through points $P, Q,$ and R. Which point is the center?

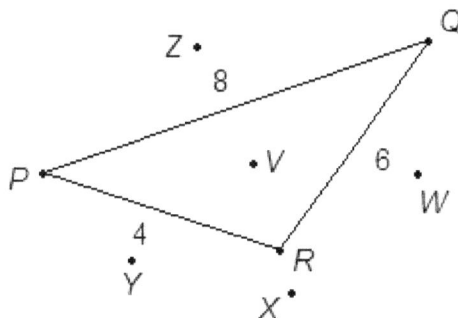

A. V
B. W
C. X
D. Y
E. Z

156. When the area of a certain circle is divided by 4π, the result is the cube of an integer. Which of the following could be the circumference of the circle?

F. 2π
G. 8π
H. 16π
J. 32π
K. 64π

157. A sphere with volume 36π cubic inches is inscribed in a cube so that the sphere touches the cube at 6 points. What is the surface area, in square inches, of the cube?

A. 216
B. 184
C. 108
D. 96
E. 36

158. In the figure below, line k has the equation $y = -x$. Line m is below line k, as shown, and m is parallel to k. Which of the following is an equation for line m ?

F. $y = -x + 3$
G. $y = -x + \sqrt{3}$
H. $y = -x - 3\sqrt{2}$
J. $y = -3x$
K. $y = -3x + 3$

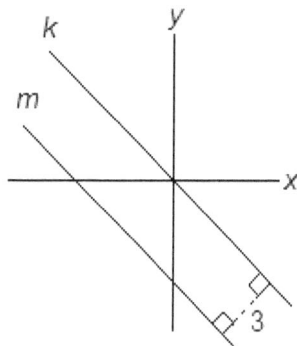

159. What is the period of the graph of $y = \frac{2}{3}\sin(\frac{5}{2}\pi\theta - 2)$?

    A. $\frac{4}{15}$

    B. $\frac{2}{5}$

    C. $\frac{4}{5}$

    D. $\frac{4\pi}{15}$

    E. $\frac{2\pi}{5}$

160. The polar equation $r\cos\theta = 2$ defines a

    F. point.
    G. circle.
    H. noncircular ellipse.
    J. line.
    K. parabola.

PROBLEMS BY LEVEL AND TOPIC
PROBLEM SET C

Full solutions to these problems are available for free download here:
www.SATPrepGet800.com/UniACTyWB

LEVEL 1: NUMBER THEORY

1. $|8 - 2| - |2 - 8| = ?$

    A. -12
    B. -4
    C. 0
    D. 4
    E. 12

2. Philip earns \$9.00 per hour for up to 40 hours of work in a week. For each hour over 40 hours of work in a week, Philip earns twice his regular pay. How much does Philip earn for a week in which he works 43 hours?

    F. \$387.00
    G. \$400.50
    H. \$414.00
    J. \$472.50
    K. \$512.00

3. What is the least integer greater than $\sqrt{67}$?

    A. 7
    B. 8
    C. 9
    D. 10
    E. 11

4. The square root of a specific number is approximately 7.6315. The specific number is between what 2 integers?

    F. 2 and 3
    G. 3 and 5
    H. 7 and 15
    J. 14 and 29
    K. 49 and 63

5. Which of the following is <u>not</u> a factor of 2431 ?

    A. 1
    B. 7
    C. 11
    D. 13
    E. 17

6. The fourth term of an arithmetic sequence is 19 and the fifth term is 13. What is the third term?

    F. -21
    G. -6
    H. 6
    J. 21
    K. 25

7. The first term is 2 in the geometric sequence $2, -4, 8, -16, \cdots$. What is the EIGHTH term of the geometric sequence?

    A. -512
    B. -256
    C. 128
    D. 256
    E. 512

8. If $(-5 + 2i) + (-1 - 3i) = a + bi$ and $i = \sqrt{-1}$, then what is the value of ab ?

    F. 6
    G. 5
    H. 4
    J. 3
    K. 2

LEVEL 1: ALGEBRA

9. If $9 + 5x = 39$, then $4x =$

    A. 6
    B. 12
    C. 18
    D. 24
    E. 30

10. For the function $f(x) = 7x^2 - 2x$, what is the value of $f(-5)$?

    F. -175
    G. -165
    H. 165
    J. 175
    K. 185

11. If $f(x) = 3(x - 1) + 5$, which of the following is equivalent to $f(x)$?

    A. $8 - 3x$
    B. $3x - 8$
    C. $3x + 2$
    D. $3x + 3$
    E. $3x + 4$

$$5y = x$$
$$5y = 70 - x$$

12. Based on the system of equations above, what is the value of x ?

    F. 10
    G. 15
    H. 25
    J. 35
    K. 40

13. The expression $9ab - 2a(4a + 5b)$ is equivalent to:

    A. $-ab - 8a^2$
    B. $2ab - 6a$
    C. $10ab - 8a^2$
    D. $-9ab$
    E. $-8a^2$

14. Which of the following expressions is equivalent to $5a + 10b + 15c$?

    F. $5(a + 2b + 3c)$
    G. $5(a + 2b + 15c)$
    H. $5(a + 10b + 15c)$
    J. $5(a + 2b) + 3c$
    K. $30(a + b + c)$

15. What is the value of the expression $(b - a)^2$ when $a = -1$ and $b = 6$?

    A. 4
    B. 9
    C. 25
    D. 36
    E. 49

16. Which of the following mathematical expressions is equivalent to the verbal expression "A number, c, squared is 52 more than the product of c and 11" ?

    F. $2c = 52 + 11c$
    G. $2c = 52c + 11c$
    H. $c^2 = 52 - 11c$
    J. $c^2 = 52 + c^{11}$
    K. $c^2 = 52 + 11c$

LEVEL 1: PROBLEM SOLVING AND DATA

17. Joe has decided to walk dogs to earn some extra money. He makes the same amount of money for each dog he walks. If he earns $360 in a week for which he walks 30 dogs, how much does he earn, in dollars, for each dog he walks?

    A. $2
    B. $4
    C. $6
    D. $12
    E. $18

18. A pack of 50 balloons is priced at $3.50 now. If the balloons go on sale for 30% off the current price, what will be the sale price of the pack?

    F. $0.45
    G. $1.75
    H. $2.00
    J. $2.45
    K. $2.50

19. The average (arithmetic mean) of five numbers is 510. If the sum of four of the numbers is 1000, what is the fifth number?

    A. 20
    B. 140
    C. 820
    D. 980
    E. 1550

20. For which of the following lists of 5 numbers is the average (arithmetic mean) less than the median?

    F. 2, 2, 4, 5, 5
    G. 2, 3, 4, 6, 7
    H. 2, 2, 4, 6, 6
    J. 2, 3, 4, 5, 6
    K. 2, 3, 4, 5, 10

21. In a jar, there are exactly 56 marbles, each of which is yellow, purple, or blue. The probability of randomly selecting a yellow marble from the jar is $\frac{2}{7}$, and the probability of randomly selecting a purple marble from the jar is $\frac{3}{7}$. How many marbles in the jar are blue?

    A. 8
    B. 12
    C. 16
    D. 20
    E. 24

22. There are exactly 50 jellybeans in a bag. There are 15 red jellybeans, 7 orange jellybeans, 13 green jellybeans, and the rest are black jellybeans. If one jellybean is selected at random from the bag, what is the probability that the jellybean is **not** black?

    F. $\frac{7}{10}$
    G. $\frac{14}{25}$
    H. $\frac{2}{5}$
    J. $\frac{3}{25}$
    K. $\frac{13}{50}$

23. A 20 member committee needs to choose a treasurer. They decide that the treasurer, who will be chosen at random, CANNOT be the president or vice president of the committee. What is the probability that Jessie, who is a member of the committee, will be chosen? (Assume that Jessie is not the president or vice president of the committee.)

    A. 0
    B. $\frac{1}{3}$
    C. $\frac{1}{9}$
    D. $\frac{1}{18}$
    E. $\frac{1}{20}$

24. If the probability that it will be cloudy tomorrow is 0.9, what is the probability that it will <u>not</u> be cloudy tomorrow?

    F. 1.4
    G. 1.0
    H. 0.6
    J. 0.1
    K. 0.0

LEVEL 1: GEOMETRY

25. Given that C is the midpoint of line segment $\overline{AB}$, $AB = 3$, $AC = x$, and $CB = 2y$, what is the value of y ?

    A. 0.5
    B. 0.75
    C. 1.25
    D. 1.5
    E. 1.75

26. In the triangle below, $x =$

    F. 62
    G. 64
    H. 66
    J. 68
    K. 70

    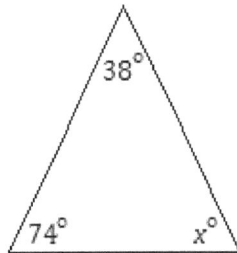

27. If the degree measures of the three angles of a triangle are $k°$, $k°$, and $85°$, what is the value of k ?

    A. 23.75
    B. 32
    C. 47.5
    D. 85
    E. 95

28. The interior dimensions of a rectangular box are 5 inches by 4 inches by 3 inches. What is the volume, in cubic inches, of the interior of the box?

    F. 12
    G. 60
    H. 90
    J. 120
    K. 124

29. What is the area of the trapezoid given in the figure below?

    A. 4
    B. 6
    C. 8
    D. 10
    E. 30

    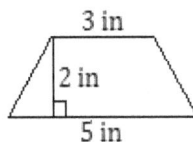

30. What is the perimeter, in inches, of a rectangle with length 15 in and width 3 in?

    F. 18
    G. 21
    H. 36
    J. 45
    K. 90

31. In the figure below, adjacent sides meet at right angles and the lengths given are in inches. What is the perimeter of the figure, in inches?

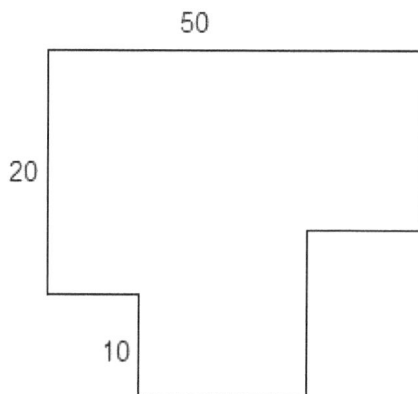

50

20

10

    A. 80
    B. 100
    C. 110
    D. 160
    E. 1500

32. What is the slope of the line through $(3, -5)$ and $(-2, -1)$ in the standard (x, y) coordinate plane?

    F. -6
    G. $-\frac{5}{4}$
    H. $-\frac{4}{5}$
    J. $\frac{4}{5}$
    K. $\frac{5}{4}$

LEVEL 2: NUMBER THEORY

33. Seven consecutive integers are listed in increasing order. If their sum is 350, what is the third integer in the list?

    A. 45
    B. 46
    C. 47
    D. 48
    E. 49

34. What is the value of $|-5| - |11 - 29|$?

    F. -45
    G. -23
    H. -13
    J. 13
    K. 23

35. Among the following rational numbers, which has the greatest value?

    A. 0.25
    B. $0.2\overline{5}$
    C. $0.\overline{25}$
    D. 0.252
    E. 0.2507

36. What is the correct ordering of 2π, 6, and $\frac{13}{2}$ from least to greatest?

    F. $6 < 2\pi < \frac{13}{2}$
    G. $2\pi < 6 < \frac{13}{2}$
    H. $2\pi < \frac{13}{2} < 6$
    J. $\frac{13}{2} < 6 < 2\pi$
    K. $6 < \frac{13}{2} < 2\pi$

37. Dana needs $5\frac{1}{12}$ ounces of a chemical for an experiment. She has $3\frac{1}{4}$ ounces of the chemical. How many more ounces does she need?

    A. $1\frac{5}{12}$
    B. $1\frac{5}{6}$
    C. $2\frac{1}{6}$
    D. $2\frac{5}{6}$
    E. $2\frac{11}{12}$

38. What is the least common multiple of 100, 70, and 30 ?

     F. 210,000
     G.   2,100
     H.   210
     J.   180
     K.   60

39. A small theatre has 12 rows of seats. The front row has 22 seats and each succeeding row has 1 less seat than the row in front of it. How many seats will be in the back row?

     A. 9
     B. 10
     C. 11
     D. 12
     E. 13

40. If $3^{6y} = 729$, then $y =$

     F. 1
     G. 2
     H. 3
     J. 4
     K. 5

LEVEL 2: ALGEBRA

41. If $\frac{13}{5}x - \frac{6}{5}x = \frac{7}{2} + \frac{7}{6}$, what is the value of x ?

     A. $\frac{3}{10}$
     B. 3
     C. $\frac{10}{3}$
     D. 10
     E. 13

42. If $\frac{1}{5} \geq \frac{7}{x}$, what is the smallest possible positive value for x ?

     F. $\frac{1}{5}$
     G. 3
     H. 17
     J. 35
     K. 70

43. A function f is defined as $f(x, y, z) = x^2 y - y^2 z + xz$. What is $f(-2, -1, 3)$?

    A. -13
    B. -7
    C. -5
    D. -1
    E. 13

$$y > 5x + 2$$
$$y - x \le 3$$

44. Which of the following ordered pairs (x, y) satisfies the system of inequalities above?

    F. $(\ 5,\ \ 2)$
    G. $(\ 0,\ \ 4)$
    H. $(-1,\ \ 1)$
    J. $(\ 1, -1)$
    K. $(\ 1,\ \ 1)$

45. $(x - 2y + 3z) - (5x - 4y + 6z)$ is equivalent to:

    A. $-6x - 6y + 9z$
    B. $-6x + 2y - 3z$
    C. $-4x - 6y - 3z$
    D. $-4x + 2y + 9z$
    E. $-4x + 2y - 3z$

46. Which of the following expressions is equivalent to $\frac{5k+50}{5}$?

    F. $k + 10$
    G. $k + 50$
    H. $7k + 10$
    J. $11k$
    K. $50k$

47. Which of the following augmented matrices represents the system of linear equations below?

$$7x + \ y = \ \ 4$$
$$5x - 3y = -2$$

    A. $\begin{bmatrix} 7 & 1 & | & -4 \\ 5 & -3 & | & 2 \end{bmatrix}$

    B. $\begin{bmatrix} 7 & 1 & | & 4 \\ 5 & -3 & | & -2 \end{bmatrix}$

    C. $\begin{bmatrix} 7 & 0 & | & 4 \\ 5 & -3 & | & -2 \end{bmatrix}$

    D. $\begin{bmatrix} 7 & -1 & | & 4 \\ 5 & -3 & | & -2 \end{bmatrix}$

    E. $\begin{bmatrix} 7 & 5 & | & 4 \\ 1 & -3 & | & -2 \end{bmatrix}$

48. For which nonnegative value of a is the expression $\frac{1}{16-a^2}$ undefined?

    F. 0
    G. 4
    H. 16
    J. 32
    K. 64

LEVEL 2: PROBLEM SOLVING AND DATA

49. Daniel is drawing a time line to represent a 1000-year period of time. If he makes the time line 75 inches long and draws it to scale, how many inches will represent 80 years?

    A. 6
    B. 9
    C. 12
    D. 20
    E. 32

50. The odometer in Don's car read 42,926 miles when he left on vacation and 43,406 miles when he returned. Don drove his car for a total of 10 hours during the trip. Based on these odometer readings, what was Don's average driving speed for the duration of the trip?

    F. 64
    G. 58
    H. 52
    J. 48
    K. 40

51. A is a set of numbers whose average (arithmetic mean) is 15. B is a set that is generated by multiplying each number in A by 6. What is the average of the numbers in set B?

    A. 15
    B. 21
    C. 45
    D. 81
    E. 90

52. What is the median of the following 9 test grades?

$$95, 72, 81, 96, 62, 98, 82, 76, 82$$

    F. 81
    G. 82
    H. 82.6
    J. 91
    K. 95

53. Jason has taken 4 tests in his math class, with grades of 73, 86, 64, and 97. In order to maintain this exact average, what *must* be Jason's grade on his 5th math test?

    A. 60
    B. 70
    C. 80
    D. 90
    E. 95

54. Five different books are to be stacked in a pile. In how many different orders can the books be placed on the stack?

    F.   5
    G.   9
    H.   15
    J.   120
    K. 3125

55. In an urn with 60 marbles, 20% of the marbles are black. If you randomly choose a marble from the urn, what is the probability that the marble chosen is <u>not</u> one of the black marbles?

    A. $\frac{1}{2}$

    B. $\frac{1}{5}$

    C. $\frac{2}{5}$

    D. $\frac{3}{5}$

    E. $\frac{4}{5}$

56. A small company released an app in early 2010. The number of downloads each year is shown in the line graph below. According to the graph, between which two consecutive years was there the greatest change in the number of app downloads?

    F. 2010 − 2011
    G. 2012 − 2013
    H. 2014 − 2015
    J. 2015 − 2016
    K. 2016 − 2017

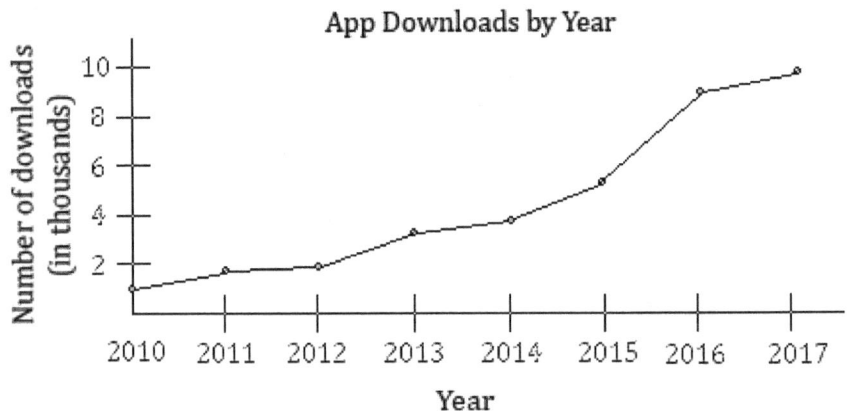

App Downloads by Year

LEVEL 2: GEOMETRY

57. If point W has a nonzero x-coordinate and a nonzero y-coordinate and the coordinates have the same sign, then point W <u>must</u> be located in which of the 4 quadrants?

    A. I only
    B. II only
    C. III only
    D. I or III only
    E. II or IV only

58. In the figure below, $\angle MAT$ measures 100°, $\angle AMT$ measures 20°, and points A, T, and H are collinear. What is the measure of $\angle MTH$?

    F. 100°
    G. 110°
    H. 120°
    J. 130°
    K. 140°

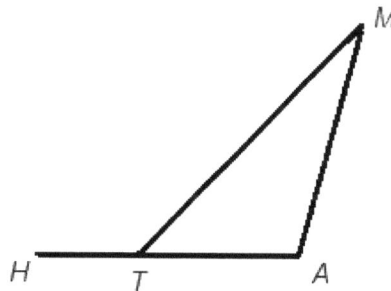

59. What is the area of a right triangle, in square centimeters, whose sides have lengths 15 cm, 36 cm, and 39 cm?

    A. 270
    B. 292.5
    C. 540
    D. 585
    E. 702

60. What is the length, in meters, of the hypotenuse of a right triangle with legs that are 5 meters long and 8 meters long, respectively?

    F. $\sqrt{89}$
    G. $\sqrt{91}$
    H. 13
    J. 20
    K. 40

61. What is the <u>diameter</u> of a circle whose circumference is 4π ?

    A. 1
    B. 2
    C. π
    D. 4
    E. 2π

62. One side of square *PQRS* is 24 feet long. A rectangle with the same area as square *PQRS* has a width of 48 feet. What is the rectangle's length, in feet?

      F.   12
      G. 16
      H. 20
      J.   48
      K. 108

63. In the right triangle pictured below, $a, b,$ and c are the lengths of its sides. What is the value of $\cos A$?

      A. $\dfrac{c}{b}$
      B. $\dfrac{a}{b}$
      C. $\dfrac{a}{c}$
      D. $\dfrac{b}{a}$
      E. $\dfrac{b}{c}$

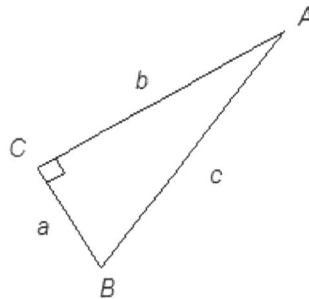

64. ΔPST is equilateral Given that $QR \parallel ST$, what is the value of x ?

      F.   60°
      G. 95°
      H. 100°
      J. 110°
      K. 120°

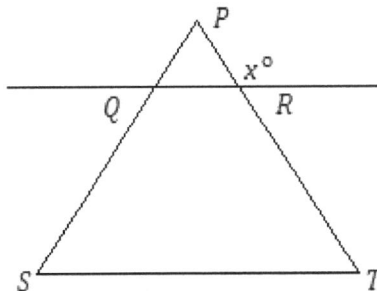

LEVEL 3: NUMBER THEORY

65. On a real number line, point X is at -3.25 and is 6.75 units from point Y. What are the possible locations of point Y on the real number line?

      A. -10 and -3.5
      B. -10 and 3.5
      C. -10 and 10
      D.   10 and -3.5
      E.   10 and 3.5

66. Which of the following is not a rational number?

    F. $\sqrt{3^2}$

    G. $\dfrac{\sqrt{\pi^2}}{\pi}$

    H. $(\sqrt{11})^2$

    J. $\sqrt{\dfrac{16}{49}}$

    K. $\sqrt{\dfrac{\pi^2}{9}}$

67. What is the largest positive integer value of k for which 7^k divides 98^{15} ?

    A. 2
    B. 7
    C. 15
    D. 28
    E. 30

68. What is the largest 2-digit integer that is divisible by 5 and is a multiple of 7 ?

    F. 21
    G. 35
    H. 70
    J. 98
    K. 105

69. It is estimated that the earth is 4.5 billion years old. When written in scientific notation, which of the following expressions is equal to the number of years used to estimate the age of the earth?

    A. 4.5×10^6
    B. 9.0×10^6
    C. 4.5×10^9
    D. 45×10^9
    E. 90×10^9

70. In the complex numbers, where $i^2 = -1$, $(4 - 3i)(4 - 3i) =$

    F. 7
    G. 25
    H. $16 - 9i$
    J. $16 + 9i$
    K. $7 - 24i$

71. Which of the following is equivalent to $8^{1/4}$?

    A. -1×8^4

    B. $\sqrt[4]{8}$

    C. $\sqrt{4}$

    D. $\frac{1}{8^4}$

    E. 4

72. If $2^x = 3$, then $2^{3x} =$

    F. 6

    G. 9

    H. 27

    J. 81

    K. 243

LEVEL 3: ALGEBRA

73. The inequality $5(x - 3) > 6(x - 2)$ is equivalent to which of the following inequalities?

    A. $x < -3$

    B. $x > -3$

    C. $x < 6$

    D. $x > 6$

    E. $x < 12$

74. The operation & is defined as $r \,\&\, s = \dfrac{s^2 - r^2}{r + s}$ where r and s are real numbers and $r \neq -s$. What is the value of $(-3) \,\&\, (-4)$?

    F. 2

    G. 1

    H. 0

    J. -1

    K. -2

75. The system of equations below has 1 solution (c, d). What is the value of c ?

$$4c + 3d = 11$$
$$-2c + 5d = 27$$

    A. -3

    B. -1

    C. 0

    D. 1

    E. 2

76. Which of the following is equivalent to the following expression?

$$16b^2 - 4a^2$$

    F. $(4b - 2a)^2$
    G. $(4b + 2a)^2$
    H. $(b - a)(16b + 4a)$
    J. $(4b - 2a)(4b + 2a)$
    K. $4b - 2a$

77. Which of the following expressions is equivalent to

$$(5x^3 - 3x^2 + 1) - (3x + 7) - (2x^2 + 1) + (5x^2 + 3x + 8) \ ?$$

    A. $5x + 1$
    B. $5x^3 + 1$
    C. $5x^3 + 10x^2 + 1$
    D. $5x^3 - 6x + 1$
    E. $5x^3 + 10x^2 - 6x + 1$

78. If $y < |x|$, which of the following is the solution statement for y when $x = -5$?

    F. $y < -5$ or $y > 5$
    G. $-5 < y < 5$
    H. $y < 5$
    J. $y > 5$
    K. y is any real number

79. It costs $(s + t)$ dollars for a box of brand A cat food, and $(q - r)$ dollars for a box of brand B cat food. The difference between the cost of 15 boxes of brand A cat food and 7 boxes of brand B cat food is k dollars. Which of the following equations represents a relationship between $s, t, q, r,$ and k ?

    A. $105(s + t)(q - r) = k$
    B. $|7(q - r) + 15(s - t)| = k$
    C. $|15(s + t) + 7(q - r)| = k$
    D. $|15(s + t) - 7(q - r)| = k$
    E. $\frac{15(s+t)}{7(q-r)} = k$

80. What value of z satisfies the matrix equation below?

$$3\begin{bmatrix} 1 & 1 & 3 \\ 2 & 2 & 0 \end{bmatrix} - 2\begin{bmatrix} 1 & 3 & 1 \\ 1 & z & 3 \end{bmatrix} = \begin{bmatrix} 1 & -3 & 7 \\ 4 & 4 & -6 \end{bmatrix}$$

    F. -2
    G. -1
    H. 0
    J. 1
    K. 2

LEVEL 3: PROBLEM SOLVING AND DATA

81. A recipe for a sports drink calls for 7 parts fruit juice to 3 parts water. To make 40 liters of this drink, how many liters of fruit juice should be used?

    A. 4
    B. 10
    C. 12
    D. 28
    E. 40

82. A room has 1700 square feet of surface that needs to be painted. If 3 gallons of paint will cover 710 square feet, what is the least whole number of gallons that must be purchased in order to have enough paint to cover the entire surface?

    F. 4
    G. 5
    H. 6
    J. 7
    K. 8

83. If 20% of a given number is 16, then what is 30% of the given number?

    A. 8
    B. 12
    C. 24
    D. 48
    E. 80

84. If the average (arithmetic mean) of a, b, and 23 is 12, what is the average of a and b ?

    F. 6.5
    G. 11
    H. 13
    J. 15
    K. It cannot be determined from the information given.

85. The average of 8 numbers is 7.3. If each of the numbers is decreased by 6, what is the average of the 8 new numbers?

    A. 0.0
    B. 0.3
    C. 1.3
    D. 2.3
    E. 7.3

86. While observing several animals in a park, John notices that the rabbit is both the 6th largest and 6th smallest animal. If every animal that John observed was a different size, how many animals did John observe?

    F. 10
    G. 11
    H. 12
    J. 13
    K. 14

87. There are y bricks in a row. If one brick is to be selected at random, the probability that it will be cracked is $\frac{3}{11}$. In terms of y, how many of the bricks are <u>not</u> cracked?

    A. $\frac{y}{11}$
    B. $\frac{8y}{11}$
    C. $\frac{11y}{8}$
    D. $\frac{3y}{11}$
    E. $11y$

88. If the letters $A, B, E, R,$ and Z are to be randomly ordered, what is the probability that the letters will appear in the order Z, E, B, R, A ?

    F. $\frac{1}{120}$
    G. $\frac{1}{60}$
    H. $\frac{1}{40}$
    J. $\frac{1}{30}$
    K. $\frac{1}{24}$

LEVEL 3: GEOMETRY

89. The lengths of two sides of right triangle $\triangle QPR$ shown below are given in centimeters. The midpoint of $\overline{PR}$ is how many centimeters from R ?

    A. 11
    B. 14
    C. 24
    D. 28
    E. 42

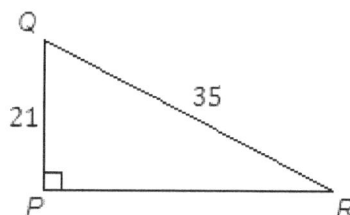

90. Square *FORM*, shown below, has side length 7 feet. What is the length, in feet, of $\overline{FR}$?

F. 49
G. 28
H. $7\sqrt{2}$
J. 7
K. $2\sqrt{7}$

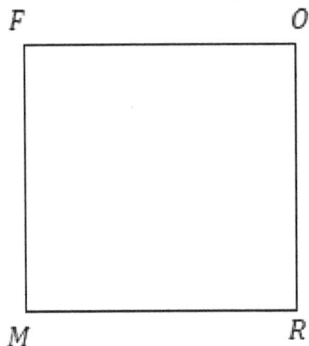

91. The volume of a right circular cylinder is 1024π cubic centimeters. If the height is twice the base radius of the cylinder, what is the base radius of the cylinder?

A. 2
B. 4
C. 6
D. 8
E. 16

92. How many figures of the size and shape below are needed to completely cover a rectangle measuring 80 inches by 30 inches?

F. 37
G. 330
H. 700
J. 740
K. 800

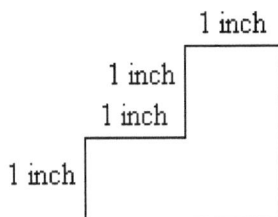

93. Which of the following equations represents the line in the standard (x, y) coordinate plane that passes through $(1, -1)$ and is perpendicular to the line $x - 3y = 1$?

A. $y = -\frac{1}{3}x + 2$
B. $y = -3x + 4$
C. $y = -3x + 2$
D. $y = 3x - 4$
E. $y = \frac{1}{3}x + 2$

94. The figure below shows a right triangle whose hypotenuse is 4 feet long. How many feet long is the shorter leg of this triangle?

 F. 2

 G. 8

 H. $2\sqrt{3}$

 J. $\dfrac{2\sqrt{3}}{3}$

 K. $\dfrac{8\sqrt{3}}{3}$

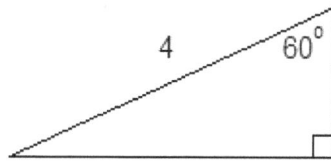

95. In the figure below, where $\Delta CAT \sim \Delta DOG$, lengths given are in inches. What is the perimeter, in inches, of ΔDOG ? (Note: The symbol $\sim$ means "is similar to.")

 A. 35
 B. 28
 C. 21
 D. 14
 E. 7

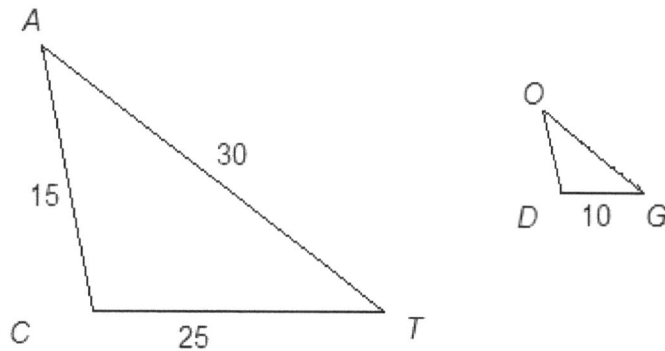

96. The 2 triangles in the figure below share a common side. What is $\cos(x + y)$?
 (Note: $\cos(x + y) = \cos x \cos y - \sin x \sin y$ for all x and y.)

 F. $\dfrac{4}{5}$

 G. $\dfrac{3 + 4\sqrt{24}}{5}$

 H. $\dfrac{4\sqrt{24} - 3}{25}$

 J. $\dfrac{3 - 4\sqrt{24}}{25}$

 K. $\dfrac{27}{25}$

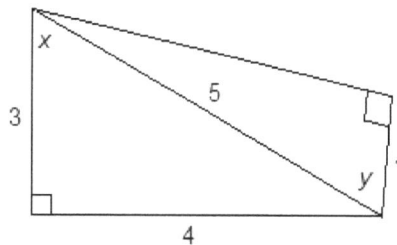

LEVEL 4: NUMBER THEORY

97. Which of the following calculations will yield twice an odd integer for any integer?

 A. $2k$
 B. $2k + 1$
 C. $4k$
 D. $4k + 1$
 E. $4k + 2$

98. Two numbers are reciprocals if their product is equal to 1. If a and b are <u>negative</u> reciprocals and $a < -1$, then b must be:

    F. greater than 1
    G. between 0 and 1
    H. equal to 0
    J. between -1 and 0
    K. less than -1

99. On Monday, Daniel received m dollars for his birthday and spent $\frac{1}{3}$ of it. On Tuesday, Daniel spent $\frac{1}{3}$ of the amount that he had left. He repeated this on Wednesday and Thursday, spending $\frac{1}{3}$ of the amount he had left each time. On Friday he spent the remaining n dollars. If m and n are both integers, what is the least possible value for m ?

    A. 81
    B. 54
    C. 27
    D. 15
    E. 12

100. The 517th digit after the decimal point in the repeating decimal $0.\overline{784398}$ is

    F. 3
    G. 4
    H. 7
    J. 8
    K. 9

101. $\frac{1}{5} \cdot \frac{3}{6} \cdot \frac{4}{7} \cdot \frac{5}{8} \cdot \frac{6}{9} \cdot \frac{7}{10} \cdot \frac{8}{11} \cdot \frac{9}{12} = ?$

    A. $\frac{1}{110}$
    B. $\frac{1}{55}$
    C. $\frac{1}{12}$
    D. 1
    E. 3

102. If $i = \sqrt{-1}$, and $\frac{(3+4i)}{(-5-2i)} = a + bi$, where a and b are real numbers, then what is the value of $29a$?

    F. 23
    G. 14
    H. 9
    J. -14
    K. -23

103. Given j and k such that $(b^3)^j = b^6$ and $(b^k)^3 = b^{12}$ for all positive b, what is b^{-k-j} ?

    A. $\dfrac{1}{b^{10}}$

    B. $\dfrac{1}{b^6}$

    C. $\dfrac{1}{b}$

    D. b^6

    E. b^{10}

104. If $\dfrac{x^a x^b}{(x^c)^d} = x^2$ for all $x \neq 0$, which of the following must be true?

    F. $\quad a + b - cd = 2$

    G. $\quad \dfrac{a+b}{cd} = 2$

    H. $\quad ab - cd = 2$

    J. $\quad ab - c^d = 2$

    K. $\dfrac{ab}{c^d} = 2$

LEVEL 4: ALGEBRA

105. The figure below shows the graph of the function f on the interval $a < x < e$. Which of the following expressions represents the difference between the maximum and minimum values of $f(x)$ on this interval?

    A. $f(b - e)$
    B. $f(b - c)$
    C. $f(a) - f(e)$
    D. $f(b) - f(c)$
    E. $f(b) - f(e)$

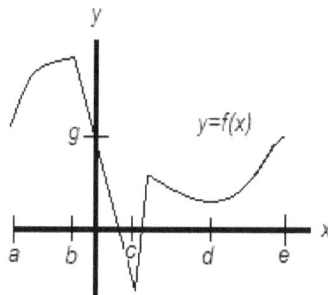

106. If $r^2 s > 10^{200}$, then the value of $\dfrac{rs + \frac{1}{r}}{7rs}$ is closest to which of the following?

    F. 0.1
    G. 0.15
    H. 0.2
    J. 0.25
    K. 0.3

107. The function g is defined by $g(x) = 4x^2 - 7$. What are all possible values of $g(x)$, where $-3 < x < 3$?

    A. $4 < g(x) < 36$
    B. $0 < g(x) < 36$
    C. $0 < g(x) < 29$
    D. $-7 \leq g(x) < 29$
    E. $-7 \leq g(x) < 0$

108. For all x, $(x^2 - 3x + 1)(x + 2) = $?

    F. $x^3 - x^2 - 5x + 2$
    G. $x^3 - x^2 - 5x - 2$
    H. $x^3 - x^2 + 5x + 2$
    J. $x^3 + x^2 + 5x + 2$
    K. $x^3 + x^2 - 5x + 2$

109. For all real numbers x and y, $|x - y|$ is equivalent to which of the following?

    A. $x + y$
    B. $\sqrt{x - y}$
    C. $(x - y)^2$
    D. $\sqrt{(x - y)^2}$
    E. $-(x - y)$

110. Tickets for a concert cost \$4.50 for children and \$12.00 for adults. 4460 concert tickets were sold for a total cost of \$29,220. Solving which of the following systems of equations yields the number of children, c, and number of adults, a, that purchased concert tickets?

    F. $c + a = 4460$
        $4.50c + 12a = 58,440$

    G. $c + a = 4460$
        $4.50c + 12a = 29,220$

    H. $c + a = 4460$
        $4.50c + 12a = 14,610$

    J. $c + a = 29,220$
        $4.50c + 12a = 4460$

    K. $c + a = 14,610$
        $4.50c + 12a = 4460$

111. The cost of 7 cupcakes is d dollars. At this rate, what is the cost, in dollars of 42 cupcakes?

    A. $\frac{6d}{7}$
    B. $\frac{d}{42}$
    C. $\frac{42}{d}$
    D. $6d$
    E. $42d$

$$-3x^2 + bx - 7$$

112. In the xy-plane, the graph of the equation above assumes its maximum value at $x = 5$. What is the value of b ?

    F. -6
    G. -2
    H. 10
    J. 20
    K. 30

LEVEL 4: PROBLEM SOLVING AND DATA

113. A mixture is made by combining a red liquid and a blue liquid so that the ratio of the red liquid to the blue liquid is 17 to 3 by weight. How many liters of the blue liquid are needed to make a 420 liter mixture?

    A. 21
    B. 42
    C. 63
    D. 147
    E. 357

114. The ratio of x to y is 5 to 1, and the ratio of y to z is 1 to 4. What is the value of $\frac{3x-2y}{5y+2z}$?

    F. 1
    G. 2
    H. 3
    J. 4
    K. 5

115. On a certain exam, the median grade for a group of 25 students is 67. If the highest grade on the exam is 90, which of the following could be the number of students that scored 67 on the exam?

        I. 5
      II. 20
     III. 24

    A. I only
    B. III only
    C. I and II only
    D. I and III only
    E. I, II, and III

116. If the average (arithmetic mean) of k and $k + 7$ is b and if the average of k and $k - 11$ is c, what is the sum of b and c ?

    F. $2k - 2$
    G. $2k - 1$
    H. $2k$
    J. $2k + \frac{1}{2}$
    K. $4k$

117. If the average (arithmetic mean) of the measures of two noncongruent angles of an isosceles triangle is 75º, which of the following is the measure of one of the angles of the triangle?

    A. 110°
    B. 120°
    C. 130°
    D. 140°
    E. 150°

118. An urn contains several marbles of which 63 are blue, 15 are red, and the remainder are white. If the probability of picking a white marble from this urn at random is $\frac{1}{3}$, how many white marbles are in the urn?

    F. 6
    G. 15
    H. 30
    J. 39
    K. 45

119. An integer from 50 through 699, inclusive, is to be chosen at random. What is the probability that the number chosen will have 2 as at least one digit?

    A. $\frac{19}{650}$
    B. $\frac{200}{651}$
    C. $\frac{4}{13}$
    D. $\frac{201}{649}$
    E. $\frac{200}{649}$

120. Jennifer has 7 shirts and 7 pairs of shoes, and each shirt matches a different pair of shoes. If she chooses one of these shirts and one pair of shoes at random, what is the probability that they will <u>not</u> match?

   F. $\dfrac{1}{7}$

   G. $\dfrac{5}{7}$

   H. $\dfrac{6}{7}$

   J. $\dfrac{1}{49}$

   K. $\dfrac{48}{49}$

LEVEL 4: GEOMETRY

121. If a square has a side of length $x + 5$ and a diagonal of length $x + 10$, what is the value of x ?

   A. 5
   B. 10
   C. 20
   D. $5\sqrt{2}$
   E. $10\sqrt{2}$

122. In the figure below, the diameters of the four semicircles are equal and lie on line segment $\overline{PQ}$. If the length of line segment $\overline{PQ}$ is $\dfrac{96}{\pi}$, what is the length of the curve from P to Q ?

   F. 96π
   G. 48π
   H. 96
   J. 48
   K. 24

123. The height of a right circular cylinder is 3 times the diameter of its base. If the volume of the cylinder is 5, what is the radius of the cylinder to the nearest hundredth?

   A. 0.32
   B. 0.64
   C. 0.76
   D. 0.85
   E. 1.26

124. The points (0,4) and (5,4) are the endpoints of one of the diagonals of a square. What is a possible y-coordinate of one of the other vertices of this square?

   F. −3.0
   G. −1.5
   H. 1.0
   J. 3.0
   K. 6.5

125. In the xy-plane, line k passes through the point $(0,-3)$ and is parallel to the line with equation $5x + 3y = 4$. If the equation of line k is $y = sx + t$, what is the value of st ?

    A. -3

    B. $-\dfrac{5}{3}$

    C. $\dfrac{5}{3}$

    D. 3

    E. 5

126. If $\cos x = 0.36$, then $\cos(180° - x) =$

    F. -0.64

    G. -0.36

    H. 0

    J. 0.36

    K. 0.64

127. In the figure below, $FROG$ is a trapezoid, Y lies on $\overleftrightarrow{FG}$, and angle measures are as marked. What is the measure of $\angle RGO$?

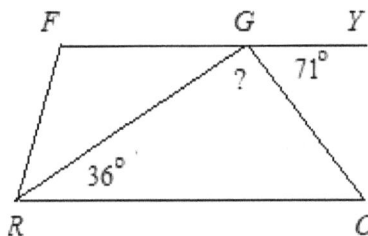

    A. $19°$

    B. $33°$

    C. $53.5°$

    D. $61°$

    E. $73°$

128. Whenever $\dfrac{\sin x}{\tan x}$ is defined, it is equivalent to:

    F. $\cos x$

    G. $\sin x$

    H. $\dfrac{1}{\cos x}$

    J. $\dfrac{1}{\sin x}$

    K. $\dfrac{1}{\cos^2 x}$

LEVEL 5: NUMBER THEORY

129. Which of the following is true for all consecutive integers j and k such that $j < k$?

    A. j is even

    B. k is even

    C. $k - j$ is even

    D. k^2 $- j$^2 is even

    E. $j^2 + k^2$ is odd

130. For all positive integers k, which of the following is a correct ordering of the terms k^{k+1}, $k^{1+k!}$, and $(k+1)^{(k+1)!}$?

    F. $k^{k+1} \geq k^{1+k!} \geq (k+1)^{(k+1)!}$
    G. $k^{1+k!} \geq k^{k+1} \geq (k+1)^{(k+1)!}$
    H. $(k+1)^{(k+1)!} \geq k^{1+k!} \geq k^{k+1}$
    J. $k^{k+1} \geq (k+1)^{(k+1)!} \geq k^{1+k!}$
    K. $k^{1+k!} \geq (k+1)^{(k+1)!} \geq k^{k+1}$

131. For every positive 3-digit number, z, with hundreds digit a, tens digit b, and units digit c, let w be the 3-digit number formed by interchanging a and c (leaving b fixed). Which of the following expressions is equivalent to $w - z$?

    A. $99(c - a)$
    B. $99(a - c)$
    C. $99c - a$
    D. $99a - c$
    E. 0

132. If $a_k = 5 + 5^2 + 5^3 + 5^4 + \cdots 5^k$, for which of the following values of k will a_k be divisible by 10 ?

    F. 5
    G. 17
    H. 66
    J. 81
    K. 99

133. The sum of an infinite geometric series with first term a and common ratio r with $-1 < r < 1$ is given by $\frac{a}{1-r}$. If the sum of a given infinite geometric series is 100 and the first term is 10. what is the value of r ?

    A. 0.1
    B. 0.3
    C. 0.5
    D. 0.7
    E. 0.9

134. The first and second terms of a geometric sequence are k and r, in that order, where k and r are positive real numbers. What is the 200th term of the sequence?

    F. $(rk)^{200}$

    G. $(rk)^{199}$

    H. $k(\frac{r}{k})^{199}$

    J. kr^{200}

    K. kr^{199}

135. The sum of the positive odd integers less than 200 is subtracted from the sum of the positive even integers less than or equal to 200. What is the resulting difference?

    A. 25
    B. 50
    C. 100
    D. 200
    E. 400

136. If $(x - 3i)(5 + yi) = 28 - 3i$, which of the following could be the value of $x + y$? (Note: $i = \sqrt{-1}$)

    F. 2
    G. 6
    H. 7
    J. 8
    K. 28

LEVEL 5: ALGEBRA

137. For all positive integers x, the function f is defined by $f(x) = (\frac{1}{b^5})^x$, where b is a constant greater than 1. Which of the following is equivalent to $f(3x)$?

    A. $\sqrt[3]{f(x)}$

    B. $(f(x))^3$

    C. $3f(x)$

    D. $\frac{1}{3}f(x)$

    E. $\frac{1}{9}f(x)$

138. Given $f(x) = \frac{2x-3}{x+5}$ and $g(x) = x^2 + 1$, which of the following is an expression for $(f \circ g)(x)$?

    F. $\frac{2x^2-1}{x^2+6}$

    G. $\frac{2x^2-1}{x+5}$

    H. $\frac{2x-3}{x^2+6}$

    J. $\frac{2x^2-2}{x^2+6}$

    K. $(\frac{2x-3}{x+5})^2 + 1$

139. If $6x = 2 + 4y$ and $7x = 3 - 3y$, what is the value of x?

    A. $\frac{9}{23}$

    B. $\frac{10}{23}$

    C. $\frac{11}{23}$

    D. $\frac{12}{23}$

    E. $\frac{13}{23}$

140. If $x + y = 2k - 1$, and $x^2 + y^2 = 9 - 4k + 2k^2$, what is xy in terms of k ?

    F. $k - 2$
    G. $(k - 2)^2$
    H. $k + 2$
    J. $(k + 2)^2$
    K. $k^2 - 4$

141. For any real number c, the equation $|x - c| = 3$ can be thought of as meaning "the distance from x to c is 3 units." How far apart are the two solutions for $|x + 2| = 3$?

    A. c
    B. $2c$
    C. $c + 3$
    D. $\sqrt{3^2 + c^2}$
    E. 6

142. On the number line, the distance between the point whose coordinate is s and the point whose coordinate is t is greater than 500. Which of the following must be true?

$$\text{I. } |s| \cdot |t| > 500$$
$$\text{II. } |s - t| > 500$$
$$\text{III. } t - s > 500$$

    F. I only
    G. II only
    H. III only
    J. I and II only
    K. I, II, and III

143. In the equation $x^2 + bx + c = 0$, b and c are integers. The only possible value for x is −5. What is the value of $b + c$?

    A. 15
    B. −15
    C. 25
    D. −25
    E. 35

144. Two numbers have a product of −75 and a sum of 0. What is the lesser of the two numbers?

    F. $-5\sqrt{3}$
    G. $-3\sqrt{5}$
    H. $-3\sqrt{2}$
    J. 0
    K. 5

LEVEL 5: PROBLEM SOLVING AND DATA

145. Tom drives an average of 150 miles each day. His car can travel an average of 22 miles per gallon of gasoline. Tom would like to reduce his daily expenditure on gasoline by $6. Assuming gasoline costs $3 per gallon, which equation can Tom use to determine how many fewer average miles, d, he should drive each day?

    A. $\frac{22}{3}d = 146$
    B. $\frac{22}{3}d = 6$
    C. $\frac{3}{22}d = 146$
    D. $\frac{3}{22}d = 6$
    E. $\frac{3}{22}d = 140$

146. An educational workshop is attended by students, teachers, tutors, administrators, and parents. Of those attending, 10% are students, 15% are teachers, 12% are tutors, 23% are administrators, and the remaining 80 people are parents. Assuming that each person in attendance has exactly one of the five roles (for example, no teacher is also an administrator), how many more teachers are in attendance than tutors?

    F. 3
    G. 4
    H. 5
    J. 6
    K. 7

147. * Jessica has two cats named Mittens and Fluffy. Last year Mittens weighed 12 pounds, and Fluffy weighed 19 pounds. Fluffy was placed on a diet, and his weight decreased by 20%. Mittens weight has increased by 20%. By what percentage did Mitten's and Fluffy's combined weight decrease, to the nearest tenth of a percent?

    A. 4.5
    B. 4.7
    C. 4.8
    D. 5.1
    E. 5.2

148. A group of students take a test and the average score is 90. One more student takes the test and receives a score of 81 decreasing the average score of the group to 87. How many students were in the initial group?

    F. 1
    G. 2
    H. 3
    J. 4
    K. 5

149. Suppose that the average (arithmetic mean) of a, b, and c is h, the average of b, c, and d is j, and the average of d and e is k. What is the average of a and e ?

    A. $h - j + k$
    B. $\dfrac{3h+3j-2k}{2}$
    C. $\dfrac{3h-3j+2k}{2}$
    D. $\dfrac{3h-3j+2k}{5}$
    E. $\dfrac{3h+3j-2k}{5}$

150. For 5 numbers in a list of increasing numbers, the average (arithmetic mean), median, and mode are all equal to 11. The range of the list is 7. The second number in the list is less than 11 and 2 more than the least number in the list. What is the greatest number in the list?

    F. 14
    G. 15
    H. 16
    J. 17
    K. 18

151. Two hundred cards numbered 100 through 299 are placed into a bag. After shaking the bag, 1 card is randomly selected from the bag. Without replacing the first card, a second card is drawn. If the first card drawn is 265, what is the probability that both cards drawn have the same tens digit?

    A. $\dfrac{19}{200}$

    B. $\dfrac{19}{199}$

    C. $\dfrac{20}{199}$

    D. $\dfrac{10}{99}$

    E. $\dfrac{24}{199}$

152. In the figure below, $ABCD$ is a square, the triangle is isosceles, $EB = 10 - 2c$, and $AD = 10$. A point in square $ABCD$ is to be chosen at random. If the probability that the point will be in the shaded triangle is $\dfrac{2}{25}$, what is the value of c ?

    F. 1
    G. 2
    H. 3
    J. 4
    K. 5

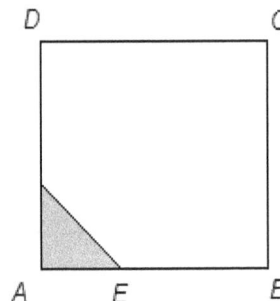

LEVEL 5: GEOMETRY

153. The lengths of the sides of a triangle are x, 16 and 31, where x is the shortest side. If the triangle is not isosceles, which of the following are possible values of x ?

> I. 14
> II. 15
> III. 16

- **A.** None
- **B.** II only
- **C.** III only
- **D.** II and III only
- **E.** I, II, and III

154. In the figure below, O is the center of the circle, the two triangles have legs of lengths a, b, c, and d, as shown, $a^2 + b^2 + c^2 + d^2 = 15$, and the area of the circle is $k\pi$. What is the value of k ?

- **F.** 7
- **G.** $\frac{15}{2}$
- **H.** 8
- **J.** $\frac{17}{2}$
- **K.** 15

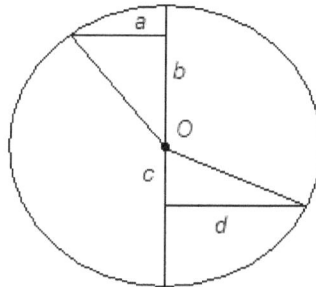

155. In the circle shown below, chords $\overline{AC}$ and $\overline{BD}$ intersect at E, which is the center of the circle. The measure of minor arc AD is 70°. What is the degree measure of $\angle BAC$?

- **A.** 15°
- **B.** 25°
- **C.** 35°
- **D.** 45°
- **E.** 55°

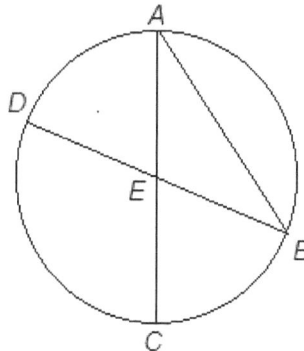

215

156. The figure shown below is a right circular cylinder. The circumference of each circular base is 20, the length of AD is 14, and AB and CD are diameters of each base respectively. If the cylinder is cut along AD, opened, and flattened, what is the length of AC ?

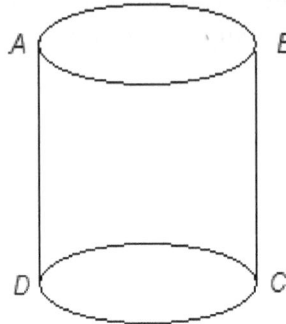

    F. $3\sqrt{3}$
    G. $\sqrt{74}$
    H. $2\sqrt{74}$
    J. $\sqrt{149}$
    K. $2\sqrt{149}$

157. Suppose that quadrilateral $PQRS$ has four congruent sides and satisfies $PQ = PR$. What is the value of $\frac{QS}{PR}$ to the nearest tenth?

    A. 1.5
    B. 1.6
    C. 1.7
    D. 1.8
    E. 1.9

158. A square, X_1, has a perimeter of 20 inches. The vertices of a second square, X_2, are the midpoints of the sides of X_1. The vertices of a third square, X, are the midpoints of the sides of X_2. This process continues indefinitely, with the vertices of X being the midpoints of the sides of X_k for each integer $k > 0$. What is the sum of the areas, in square in., of $X_1, X_2, \dots$?

    F. $\frac{20}{3}$

    G. 20

    H. 35

    J. 50

    K. 100

159. If $\tan \theta = -\frac{3}{4}$ and $\frac{3\pi}{2} < \theta < 2\pi$, then $\cos \theta = $?

    A. $-\frac{5}{4}$
    B. $-\frac{4}{5}$
    C. $-\frac{3}{5}$
    D. $\frac{3}{5}$
    E. $\frac{4}{5}$

160. In triangle PQR, the measure of $\angle P$ is $x°$, $PQ = a$ ft, and $PR = b$ ft. Which of the following is the length, in feet, of $\overline{QR}$?

    F. $a \sin 32°$
    G. $b \sin 32°$
    H. $\sqrt{b^2 - a^2}$
    J. $\sqrt{b^2 + a^2}$
    K. $\sqrt{b^2 + a^2 - 2ba \cos x°}$

UNI SAYS...

The Challenge Problems on the following page are NOT ACT questions. However, working out the solutions to these problems WILL increase your level of mathematical maturity. If you are trying to get a perfect score in ACT math, but cannot seem to get past the low 30's, then a boost in your mathematical skill level may be exactly what you need to break through this final sticking point. Make sure to spend some time struggling with each problem before checking the solution. It is the struggle that will make your mind stronger. Best of luck!

Full solutions to the Challenge Problems are available for free download here:
www.SATPrepGet800.com/UniACTyWB

CHALLENGE PROBLEMS

1. The integer n is equal to k^3 for some integer k. Suppose that n is divisible by 45 and 400. What is the smallest possible value of n ?

2. If a and b are positive integers, $\left(a^{\frac{1}{2}}b^{\frac{1}{3}}c^{\frac{1}{5}}\right)^{30} = 41,472$, and $abc = 1$, what is the value of ab ?

3. Find the smallest positive integer k such that $\frac{k}{2}$ is a perfect square and $\frac{k}{3}$ is a perfect cube.

4. Let f be a linear function such that $f(5) = -2$ and $f(11) = 28$. What is the value of $\frac{f(9)-f(7)}{2}$?

5. Given a list of 50 positive integers, each no bigger than 98, show that at least one integer in the list is divisible by another integer in the list.

6. Show that $\sqrt{5} - \sqrt{9 - 4\sqrt{5}}$ is an integer.

7. Let j and k be positive integers with $j \le k$. In how many ways can k be written as a sum of j positive integers?

8. How many arrangements of the word EFFLORESCENCE have consecutive C's and F's but no consecutive E's?

9. Find the sum of the integers from 24 to 276 inclusive.

10. Show that the sum of an arithmetic series is $A_n = n \cdot m$ where n is the number of terms and m is the average (arithmetic mean) of the first and last term.

11. Draw a rectangular solid with sides of length a, b and c, and let the long diagonal have length d. Show geometrically that $d^2 = a^2 + b^2 + c^2$.

12. If 2 real numbers are randomly chosen from a line segment of length 10, what is the probability that the distance between them is at least 7 ?

13. Use the formula $d = rt$ to derive Xiggi's formula.

14. The graphs of $y = bx^2$ and $y = k - bx^2$ intersect at points A and B. If the length of $\overline{AB}$ is equal to d, what is the value of $\frac{bd^2}{k}$?

15. Show that if x is the least integer in a set of $n + 1$ consecutive integers, then the median of the set is $x + \frac{n}{2}$.

16. Show that in a set of consecutive integers, the average (arithmetic mean) and median are equal.

ACTIONS TO COMPLETE AFTER YOU HAVE READ THIS BOOK

1. Take another practice ACT

You should see a substantial improvement in your score.

2. Continue to practice ACT math problems for 10 to 20 minutes each day

3. 'Like' my Facebook page

This page is updated regularly with SAT prep advice, tips, tricks, strategies, and practice problems. Visit the following webpage and click the 'like' button.

www.facebook.com/Get800

4. Review this book

If this book helped you, please post your positive feedback on the site you purchased it from; e.g. Amazon, Barnes and Noble, etc.

5. Claim your FREE bonus

If you have not done so yet, visit the following webpage and enter your email address to receive solutions to all the problems in this book.

www.SATPrepGet800.com/UniACTyWB

About the Author

Dr. Steve Warner, a New York native, earned his Ph.D. at Rutgers University in Pure Mathematics in May 2001. While a graduate student, Dr. Warner won the TA Teaching Excellence Award.

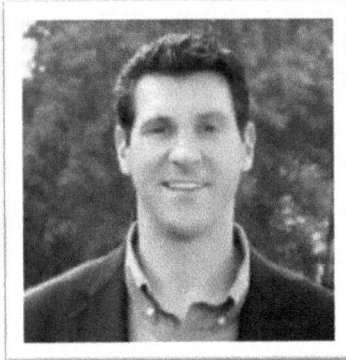

After Rutgers, Dr. Warner joined the Penn State Mathematics Department as an Assistant Professor. In September 2002, Dr. Warner returned to New York to accept an Assistant Professor position at Hofstra University. By September 2007, Dr. Warner had received tenure and was promoted to Associate Professor. He has taught undergraduate and graduate courses in Precalculus, Calculus, Linear Algebra, Differential Equations, Mathematical Logic, Set Theory and Abstract Algebra.

Over that time, Dr. Warner participated in a five-year NSF grant, "The MSTP Project," to study and improve mathematics and science curriculum in poorly performing junior high schools. He also published several articles in scholarly journals, specifically on Mathematical Logic.

Dr. Warner has more than 15 years of experience in general math tutoring and tutoring for standardized tests such as the SAT, ACT and AP Calculus exams. He has tutored students both individually and in group settings.

In February 2010 Dr. Warner released his first SAT prep book "The 32 Most Effective SAT Math Strategies," and in 2012 founded Get 800 Test Prep. Since then Dr. Warner has written books for the SAT, ACT, SAT Math Subject Tests, AP Calculus exams, and GRE.

Dr. Steve Warner can be reached at

steve@SATPrepGet800.com

BOOKS BY DR. STEVE WARNER

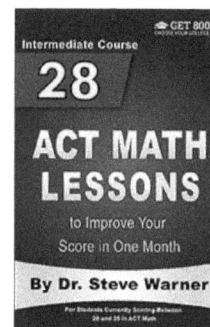

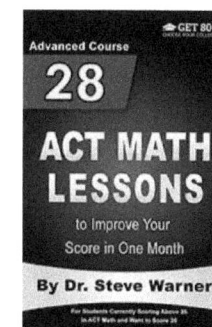